Global Issues Series

General Editor: **Jim Whitman**

This exciting new series encompasses three principal themes: the interaction of human and natural systems; cooperation and conflict; and the enactment of values. The series as a whole places an emphasis on the examination of complex systems and causal relations in political decision-making; problems of knowledge; authority, control and accountability in issues of scale; and the reconciliation of conflicting values and competing claims. Throughout the series the concentration is on an integration of existing disciplines towards the clarification of political possibility as well as impending crises.

Titles include:

Berhanykun Andemicael and John Mathiason
ELIMINATING WEAPONS OF MASS DESTRUCTION
Prospects for Effective International Verification

Robert Boardman
GOVERNANCE OF EARTH SYSTEMS
Science and Its Uses

Mike Bourne
ARMING CONFLICT
The Proliferation of Small Arms

John N. Clarke and Geoffrey R. Edwards (*editors*)
GLOBAL GOVERNANCE IN THE TWENTY-FIRST CENTURY

Neil Davison
'NON-LETHAL' WEAPONS

Nicole Deitelhoff and Klaus Dieter Wolf (*editors*)
CORPORATE SECURITY RESPONSIBILITY?
Corporate Governance Contributions to Peace and Security in Zones of Conflict

Toni Erskine (*editors*)
CAN INSTITUTIONS HAVE RESPONSIBILITIES?
Collective Moral Agency and International Relations

Moira Feil
GLOBAL GOVERNANCE AND CORPORATE RESPONSIBILITY IN CONFLICT ZONES

Annegret Flohr, Lothar Rieth, Sandra Schwindenhammer and Klaus Dieter Wolf
THE ROLE OF BUSINESS IN GLOBAL GOVERNANCE
Corporations as Norm-Entrepreneurs

Annegret Flohr
SELF-REGULATION AND LEGALIZATION
Making Global Rules for Banks and Corporations

Beth K. Greener
THE NEW INTERNATIONAL POLICING

David Karp and Kurt Mills (*editors*)
HUMAN RIGHTS PROTECTION IN GLOBAL POLITICS
The Responsibilities of States and Non-State Actors

Alexander Kelle, Kathryn Nixdorff and Malcolm Dando
CONTROLLING BIOCHEMICAL WEAPONS
Adapting Multilateral Arms Control for the 21st Century

Kelley Lee
HEALTH IMPACTS OF GLOBALIZATION (*editor*)
Towards Global Governance

Kelley Lee
GLOBALIZATION AND HEALTH
An Introduction

Catherine Lu
JUST AND UNJUST INTERVENTIONS IN WORLD POLITICS
Public and Private

Robert L. Ostergard Jr. (*editor*)
HIV, AIDS AND THE THREAT TO NATIONAL AND INTERNATIONAL SECURITY

Graham S. Pearson
THE UNSCOM SAGA
Chemical and Biological Weapons Non-Proliferation

Graham S. Pearson
THE SEARCH FOR IRAQ'S WEAPONS OF MASS DESTRUCTION
Inspection, Verification and Non-Proliferation

Nick Ritchie
A NUCLEAR WEAPONS-FREE WORLD?
Britain, Trident and the Challenges Ahead

Julian Schofield
STRATEGIC NUCLEAR SHARING

David Scott
'THE CHINESE CENTURY'?
The Challenge to Global Order

Andrew Taylor
STATE FAILURE

Marco Verweij and Michael Thompson (*editors*)
CLUMSY SOLUTIONS FOR A COMPLEX WORLD
Governance, Politics and Plural Perceptions

Marco Verweij
CLUMSY SOLUTIONS FOR A WICKED WORLD
How to Improve Global Governance

Global Issues Series
Series Standing Order ISBN 978–0–333–79483–8
(*outside North America only*)

You can receive future titles in this series as they are published by placing a standing order. Please contact your bookseller or, in case of difficulty, write to us at the address below with your name and address, the title of the series and the ISBN quoted above.

Customer Services Department, Macmillan Distribution Ltd, Houndmills, Basingstoke, Hampshire RG21 6XS, England

Neuroscience and the Future of Chemical-Biological Weapons

Malcolm Dando
Professor of International Security, Department of Peace Studies, University of Bradford, UK

palgrave
macmillan

First published 2015 by
PALGRAVE MACMILLAN

Palgrave Macmillan in the UK is an imprint of Macmillan Publishers Limited, registered in England, company number 785998, of Houndmills, Basingstoke, Hampshire RG21 6XS.

Palgrave Macmillan in the US is a division of St Martin's Press LLC, 175 Fifth Avenue, New York, NY 10010.

Palgrave Macmillan is the global academic imprint of the above companies and has companies and representatives throughout the world.

Palgrave® and Macmillan® are registered trademarks in the United States, the United Kingdom, Europe and other countries.

ISBN: 978–1–137–38181–1

This book is printed on paper suitable for recycling and made from fully managed and sustained forest sources. Logging, pulping and manufacturing processes are expected to conform to the environmental regulations of the country of origin.

A catalogue record for this book is available from the British Library.

A catalog record for this book is available from the Library of Congress.

Contents

List of Figures

List of Tables

Preface and Acknowledgements

Although I trained originally as a biologist I have worked on arms control and disarmament issues for the last 35 years and since the mid-1990s my work has been focused on the problem of strengthening biological[1] and chemical[2] non-proliferation regimes. Thus I followed the decade-long effort to add a more adequate verification system to the Biological and Toxin Weapons Convention (BTWC) during the 1990s and early years of this century.[3] When these efforts came to an end in the chaos of the 2001–2002 Fifth Review Conference and codes of conduct for life scientists became part of the agenda for the 2005 intersessional meetings, I worked with Brian Rappert of the University of Exeter to try to discover what practising life scientists thought about biosecurity and dual-use issues. We held meetings with numerous scientists in 16 countries[4] and, to my astonishment then but not now, we found that very few of them had even heard of the BTWC or the more recently agreed Chemical Weapons Convention (CWC), let alone about the biosecurity and dual-use issues we wished to discuss with them.

Subsequently, working with colleagues at Bradford and elsewhere, I investigated whether the reason for this gap in their knowledge was that these conventions, and the responsibilities of scientists in relation to the BTWC and the CWC, were not covered in the university education of life scientists. Again, in a number of countries we found that this was indeed the situation, very few courses for life scientists covered biosecurity and dual-use issues.[5,6,7,8] This led us to work with colleagues at Japan's National Defense Medical College to produce an online Education Module Resource (EMR) to help lecturers teaching life scientists to incorporate into their courses material from the EMR that they thought appropriate. The EMR has subsequently been translated into a number of languages in addition to the original Japanese and English.[9]

Given that almost all the public discussions of dual-use have involved experiments in microbiology, it is perhaps not surprising that most efforts at raising awareness and developing educational material on biosecurity and dual-use for life scientists have focused on this area of the life sciences – from the mousepox experiment[10] through to the current concerns about gain-of-function (GOF) experiments with deadly influenza viruses.[11] However, it has been clear since the Lemon–Relman[12] report for the US National Academies almost a decade ago that this problem of dual-use – the danger that the results of benignly intended work in the life sciences would later be misused for hostile purposes by others – ranges far wider than just microbiology.

As the history of large-scale offensive chemical and biological weapons programmes carried out by major states in the last century shows, one field where advances could obviously be of concern in the future is neuroscience. It is only necessary to recall the recent use of nerve agents in Syria to see how the discovery of acetylcholine chemical neurotransmission was followed quickly by the finding, in civil work, of the means of disrupting such neurotransmission in pest control and then the translation of that knowledge into a major increase in the toxicity of chemical warfare agents.

For that reason I have recently been involved with colleagues at the University of Manchester and elsewhere in the UK in trying to construct and implement educational material on biosecurity and dual-use for practising neuroscientists.[13] We see this as an initial example of what will be necessary to engage many scientists in other fields of the life and associated sciences as it is quite clear that plant pathologists and veterinarians, for example, will require rather different educational content to be produced in order to engage their interest effectively.

This book is targeted at practising neuroscientists, but it acknowledges that they can come from very diverse backgrounds, including molecular biology, information technology or pharmacology. Therefore, it is not assumed that matters that would be simple to a neurophysiologist, for example, do not need to be briefly explained. This is also a necessary approach as another target audience for this book is international relations and international security scholars who are increasingly concerned that we take better care in protecting the benignly intended and extremely useful results of work in the life sciences from misuse in this conflict-ridden early twenty-first century.

I have tried to make a complex subject simpler to follow by having a roughly chronological framework of three parts: the past, the present and the future. While avoiding going too deeply into technical detail I have tried also to give sufficient references so that any particular aspect can be followed at the technical level if that is considered useful by a reader.

In Part I, Chapter 1 outlines how and why the BTWC and CWC were negotiated and why neuroscience research was, and still is, relevant to the development of chemical and biological weapons. Chapter 2 provides a brief overview of the aspects of the nervous system that are most relevant to thinking about weapons attacking the brain. Chapter 3 describes the development of the Chemical and Biological Weapons (CBW) Non-Proliferation Regime and Chapter 4 deals with the challenges posed to that regime by research that could be dual-use.

In Part II, Chapter 5 gives an insight into modern civil neuroscience by describing the initial stages of the recently announced United States and European Union brain research projects. Chapter 6 uses a series of recent reports to show how concerns about novel neuroweapons based on advances in neuroscience (and related sciences and technologies) have come about. Chapter 7 attempts to put these concerns in a wider long-term perspective by examining what is being discovered in the nascent field of neuroparisitology about how parasites have evolved mechanisms to manipulate specific neuronal circuits in their unfortunate hosts in order to produce behaviour that benefits the parasite. Chapters 8 and 9 then deal with aspects of modern neuroscience that could most easily be misused, in particular incapacitating chemical agents that could be seen to have utility in the 'wars amongst the people' that characterise current conflicts.

In Part III Chapter 10 reviews how those states party to the BTWC and the CWC have tried to deal with the problem of rapid scientific and technological change in recent years. Chapter 11 seeks to describe where we are at present in international efforts to prevent the future misuse of advances in neuroscience and argues that we still have the opportunity to put policies in place that will minimise the possibility of such misuse. Chapter 12 concludes by reviewing what needs to be done now in that regard, focusing particularly on how the awareness and education of practising neuroscientists might best be improved for them to engage their expertise in the process of developing sensible policies that will not overly restrict research while significantly hindering possible misuse.

Parts of the research for this book have been funded by grants to Professor Brian Rappert at the University of Exeter (ES/K011308/1, The formulation and non-formulation of security concerns) and Professor David Galbreath at the University of Bath (ES/K011227/1, Biochemical security 2030), and the figures in Chapter 2 have been redrawn from my original diagrams.[14]

References

1. Dando, M. R. (1994) *Biological Warfare in the 21st Century: Biotechnology and the Proliferation of Biological Weapons*. London: Brassey's.
2. Dando, M. R. (1996) *A New Form of Warfare: The Rise of Non-Lethal Weapons*. London: Brassey's.
3. Dando, M. R. (2002) *Preventing Biological Warfare: The Failure of American Leadership*. Basingstoke: Palgrave Macmillan.
4. Rappert, B. (2009) *Experimental Secrets: International Security, Codes and the Future of Research*. Lanham: University Press of America.
5. Mancini, G. and Revill, J. (2008) *Fostering the Biosecurity Norm: Biosecurity Education for the Next Generation of Life Scientists*. Landau Network-Centro Volta and University of Bradford. Available at http://www.dual-usebioethics.net.
6. Minehata, M. and Shinomiya, N. (2009) *Biosecurity Education: Enhancing Ethics, Securing Life and Promoting Science: Dual-Use Education in Life Science Courses in Universities in Japan*. Saitama, Japan and University of Bradford. Available at http://www.dual-usebioethics.net.
7. Minehata, M. and Friedman, D. (2010) *Biosecurity Education in Israel Research Universities: Survey Report*. Institute for National Security Studies and University of Bradford. Available at http://www.dual-usebioethics.net.
8. Bartolucci, V. and Dando, M. R. (2011) What does neuroethics have to say about the problem of dual-use? Pp. 27–42 in B. Rappert and M. J. Selgelid, *On the Dual Uses of Science and Ethics: Principles, Practices and Prospects*. Canberra: Australian National University EPress.
9. The *Education Module Resource* is available in various languages at www.dual-usebioethics.net.
10. Shinomiya, N., Minehata, M. and Dando, M. R. (2013) Bioweapons and dual-use research of concern. *Journal of Disaster Research*, **8** (4), 654–666.
11. Wain-Hobson, S. (2013) H5N1 viral-engineering dangers will not go away. *Nature*, 495, 411.
12. National Academies (2006) *Globalization, Biosecurity, and the Future of the Life Sciences*. Washington, D.C.: National Academies Press
13. See http://www.lab.ls.manchester.ac.uk/neuroethicseducation/educational-module/
14. See reference 2.

List of Abbreviations

AAAS	American Association for the Advancement of Science
ACh	Acetylcholine
AG	Australia Group
BBB	Blood–Brain Barrier
BRAIN	Brain Research through Advancing Innovative Neurotechnologies
BTWC	Biological and Toxin Weapons Convention
BZ	3-Quinuclidinyl Benzilate
CBM	Confidence-Building Measure
CBRN	Chemical, Biological, Radiological and Nuclear
CBW	Chemical and Biological Weapons
CNS	Central Nervous System
CPG	Central Pattern Generator
CWC	Chemical Weapons Convention
DA	Dopamine
DARPA	Defense Advanced Research Projects Agency (United States)
DC	Dendritic Cells (of the immune system)
DG	Dorsal Dentate Gyrus (of the hippocampus)
DNA	Deoxyribonucleic Acid
L-DOPA	L-3,4-Dihydroxyphenylalanine (neurotransmitter precursor)
DURC	Dual-use Research of Concern
EEE	Eastern Equine Encephalitis
ELSI	Ethical, Legal, Social Implications
EMR	Education Module Resource
EQuATox	A Network of Expert Labs Organising Tests to Detect Toxins
EU	European Union
F4/AB45Y-4	US Incapacitant Weapon System
FMRI	Functional Magnetic Resonance Imaging
FVR	Foundation for Vaccine Research
GA	Tabun
GABA	Gamma-aminobutyric Acid
GB	Sarin
GD	Soman

GOF	Gain-of-Function (experiments)
HBP	Human Brain Project
IAEA	International Atomic Energy Agency
IAP	Inter-Academy Panel
IAU	Investigation of Alleged Use
ICA	Incapacitating Chemical Agent
INSEN	International Nuclear Security Education Network
ISP	Intersessional Process
ISU	Implementation Support Unit
IUPAC	International Union of Pure and Applied Chemistry
LC	Locus Coeruleus
LSD	Lysergic Acid Diethylamide
MERS	Middle East Respiratory Syndrome
NA	Noradrenaline
NE	Norepinephrine
NIH	National Institutes of Health (United States)
NSABB	National Science Advisory Board for Biosecurity (United States)
NSID	National security, intelligence and defence
OA	Octopamine
OCPF	Other Chemical Production Facility
OPCW	Organization for the Prohibition of Chemical Weapons
PG	Staphylococcal Enterotoxin B (weaponised agent)
PET	Positron Emission Tomography
PTSD	Post-traumatic Stress Disorder
RCA	Riot Control Agent
RCR	Responsible Conduct of Research
REM	Rapid Eye Movement (sleep)
SAB	Scientific Advisory Board (of the OPCW)
SAI	Standing Agenda Item
SARS	Severe Acute Respiratory Syndrome
SEB	Staphylococcal Enterotoxin B
SEG	Sub-oesophageal Ganglion
SIPRI	Stockholm International Peace Research Institute
SUBNETS	Systems-based Neurotechnology for Emerging Therapies
SupEG	Supra-oesophageal Ganglion
SWORDS	First Generation Military Robot
TS	Technical Secretariat (of the OPCW)
TWG	Temporary Working Group (of the SAB)
UCSF	University of California at San Francisco
UK	United Kingdom of Great Britain and Northern Ireland

US	United States of America
USG	United States Government
UV	Ultraviolet (light)
V	Series of Nerve Agents (such as VX)
VEE	Venezuelan Equine Encephalitis
VEREX	Committee of the BTWC researching verification measures
VLPO	Ventrolateral Preoptic (area of the brain)
WEE	Western Equine Encephalitis
WHO	World Health Organization

Part I

The Past

1
Neuroscience and CBW

Introduction

States are unlikely to spend the time and effort required to negotiate and implement international multilateral arms control and disarmament agreements unless there are serious problems that require the use of such complex methods. Multilateral negotiations can take years to conclude, resulting in the need for extensive national implementation and ongoing multilateral engagement in order to assess the operation of the agreement and how it might need further elaboration.

Therefore, the fact that over the last century three such international multilateral agreements were negotiated and implemented in relation to the control of chemical and biological weapons leaves little doubt that many states perceived that such weapons were a significant threat. During the terrible war-torn twentieth century[1] the 1925 Geneva Protocol, the 1975 Biological and Toxin Weapons Convention (BTWC) and the 1997 Chemical Weapons Convention (CWC) progressively brought tighter and tighter control over the proliferation of these weapons.

The 1925 Geneva Protocol was negotiated following the large-scale use of chemical weapons, and the initial crude attempts to use (anti-animal) biological weapons,[2] during the First World War. Now that most of the many reservations that were lodged at the time have been removed, the Protocol bans the use of chemical and biological weapons.[3] In 2012 the United Nations General Assembly once again reaffirmed the 'vital necessity' that states uphold the provisions of the Protocol and called upon States still holding reservations to withdraw them.[4]

The Biological and Toxin Weapons Convention was opened for signature in 1972 and entered into force in 1975. Its first article adds a series of further prohibitions to the ban on use stating, in part, that:[5]

> Each State Party to this Convention undertakes never in any circumstances to develop, produce, stockpile or otherwise acquire or retain:
>
> 1. Microbial or other biological agents, or toxins whatever their origin or method of production, of types and in quantities that have no justification for prophylactic, protective or other peaceful purposes.

Thus, under what has become known as the General Purpose Criterion any peaceful uses of biological and toxin agents are allowed but non-peaceful purposes are banned. Like the 1925 Geneva Protocol, the BTWC continues to be developed through its five-yearly Review Conferences, the latest of which took place in 2011.[6]

Whilst Article IX[7] of the BTWC recognised the 'objective of effective prohibition of chemical weapons' and states party to it undertook 'to continue negotiations in good faith with a view to reaching early agreement', it was not until the end of the East-West Cold War that the Chemical Weapons Convention was agreed. It opened for signature in 1993 and entered into force in 1997.

Article I of the CWC states, in part, that[8]

> 1. Each State Party to this Convention undertakes never under any circumstances:
>
> (a) To develop, produce, otherwise acquire, stockpile or retain chemical weapons, or transfer, directly or indirectly chemical weapons to anyone;

The CWC clearly also has a General Purpose Criterion applying to all chemicals as Article II defines a chemical weapon, in part, as:

> (a) Toxic chemicals and their precursors, except where intended for purposes not prohibited under this Convention, as long as the types and quantities are consistent with such purposes.

The CWC is also subject to a review every five years and the latest such review took place in 2013.[9]

The responsible conduct of research

Why should a practising neuroscientist carrying out benignly intended civil research be interested in such international arms control issues? Surely, it might well be argued, there is enough to do keeping up with this rapidly advancing field and making a reasonable research contribution in his or her own area of cutting-edge research. That, however, would be to ignore the evolution of the scientific community's conception of responsible conduct of research (RCR). As Rebecca Carlson and Mark Frankel of the American Association for the Advancement of Science (AAAS) explained recently, the scientific community is getting better at teaching and observing the necessities of responsible conduct in regard to the internal operations of science, such as dealing with human and animal subjects, data acquisition and its management, and publication practices and responsibilities.[10] However, they also argue that there is a long way to go before the scientific community can be said to have dealt adequately with its external research responsibilities. These cover aspects of the societal impacts of research, such as communication, advocacy and emerging technologies.

Two questions that really need to be asked are these. What evidence is there that, as the growing sciences of chemistry and biology have been applied to the development and use of chemical and biological weapons over the last one hundred years, advances in neuroscience have contributed to these hostile purposes? And what are the possibilities that such distortions of civil science might continue in the future? One way to approach those questions is to look at the weapon systems that have been produced by states and the extent to which they affect the nervous system.

The nervous system is made up of individual cellular units, including the numerous neurons that are specialised for the transmission of information to, from and within the central nervous system. Information transmission within a neuron is by electrical means, but transmission between neurons is predominantly by chemical means. Specialised neurotransmitter molecules are released from one neuron into the synaptic cleft and latch onto specific receptor molecules on the next cell in order to affect the operation of that following neuron or effector system (such as muscle). This, of course, opens up the possibility of manipulation of the nervous system by the introduction of other chemicals, like drugs for benign purposes or chemical agents for hostile purposes. It should be understood, however, that the nervous system does not act in isolation and is intimately linked to the endocrine (hormonal) system and the immune (defence) system.[11] Thus, stress registered in the brain can lead to hormones being released that then cause the release of glycogen and glucose, readily available substrates for energy metabolism, whilst amazingly small amounts of some bacterial toxins can induce elevation of body temperature in the fever response to infection. Given that there are many different neurotransmitters, hormones and cytokines (of the immune system) and numerous cellular receptors for bioregulatory chemicals in the nervous system, it follows that as our knowledge becomes more detailed, more and more specific targets for manipulation – for benign or malign purposes – are likely to be revealed.

Chemical weapons

Michael Faraday is, of course, best known for his groundbreaking work on electricity in the first half of the nineteenth century. It is less well known that Faraday was also a significant chemist. He was, for example, the discoverer of benzene, which was to be of fundamental importance in the growth of organic chemistry later in the century.[12] In later life Faraday was frequently consulted by officialdom for his views on scientific issues and during the Crimean War he was asked for his opinion on a proposed scheme to attack and capture Cronstadt through the use of a chemical weapon. As his biographer[13] noted, 'Faraday was sceptical of the plan and his report could not be interpreted as a favourable one'. Indeed, it was not until the First World War, after the growth of industrial chemistry in the latter part of the nineteenth century, that large-scale chemical warfare became possible.

Chemical weapons can reasonably be divided into lethal and disabling agents[14] (Table 1.1). Lethal agents such as phosgene and mustard gas were used in large quantities in the First World War, but it was not until the 1930s that nerve agents were first discovered, in Germany. Disabling incapacitating agents like BZ that affect the central nervous system, were developed after the Second World War as drugs began to be discovered that could help people suffering from some mental illnesses.

Acetylcholine (ACh), the first neurotransmitter molecule to be discovered, resulted from research by Loewi early in the twentieth century.[15] ACh is manufactured in some neurons and stored in vesicles on the presynaptic side of the synaptic cleft between an ACh neuron and a

Table 1.1 Chemical weapons agents

Lethal	Lung irritants	e.g. Phosgene
	Blood gases	e.g. Hydrogen cyanide
	Vesicants	e.g. Mustard gas
	Nerve gases	e.g. Sarin, VX
Disabling	Incapacitants	e.g. LSD, Agent BZ
	Harassing agents and other irritants	e.g. Agent CN, Agent CS, Agent OC

postsynaptic neuron. When a nerve impulse (an electrical signal) in the presynaptic neuron reaches the synapse the ACh is released into the cleft, attaches to receptors on the postsynaptic neuron, and affects the electrical activity of that cell. However, precision in the information transfer is ensured because an enzyme called acetylcholinesterase quickly breaks down the ACh in the synaptic cleft. The constituent parts of the ACh molecule are then taken up for reuse in the presynaptic neuron.[16]

Nerve agents are deadly because their main action is to inhibit the action of acetylcholinesterase and thus excessive amounts of ACh accumulate in the synaptic cleft and continue to affect postsynaptic cells. As there are ACh synapses in the skeletal muscles, the autonomic nervous system and the brain, it is no surprise that agents such as GA (tabun), GB (sarin), GD (soman) and the even more toxic V agents cause extensive disruption of bodily functions and can lead to death.[17] The original G series of nerve agents were discovered by civil scientists working on pesticides in Germany before the Second World War[18] and then, as shown clearly in a recent study, the V agents were discovered through later civil research after that war.[19]

The example of the initial development of the V agents is of particular interest because of the involvement of UK civil scientists. As the authors explain:[20]

> Although defence research and development laboratories achieved incremental improvements in chemical warfare agents, major breakthroughs such as the discovery of the G [original series] and V-agents were spin-offs of civil technologies.

They go on to explain how the transfer of Amiton from Plant Protection Limited to the Chemical Defence Experimental Establishment at Porton Down demonstrated the link between the civil industry and the defence establishment. We will have cause to return to this point, that of crucial breakthroughs in weapons developments resulting from *civil* not military research. It should be noted that even such dangerous agents as sarin, once developed for military purposes, were eventually produced and used by terrorists in the Tokyo subway attack of 1995.[21]

It should also be noted that the lethal chemical agents used during the First World War could have effects on the central nervous system. As the US *Textbook of Military Medicine*[22] notes, although 'the effects are not usually prominent clinically, mustard affects the CNS [central nervous system]'. The account goes on to say that large amounts of mustard gas administered by various routes to animals caused 'convulsions, and other neurological manifestations' and that they died a 'neurological death' a few hours after being given a lethal dose.

The effect of a chemical agent is a function of dose, so not everyone will be killed by the release of a nerve agent or other lethal chemical, but the World Health Organization (WHO) defines lethal chemicals[23] as those 'intended either to kill or injure the enemy so severely as to necessitate evacuation and medical treatment'. On the other hand, it defines disabling chemicals as those 'used to incapacitate the enemy by causing a disability from which recovery may be possible without medical aid'. The WHO also points out that, when the industrial developments of the nineteenth century allowed the large-scale use of chemical weapons in the First World War, chemical warfare began with the use of sensory irritants, such as tear gases, mainly to drive enemy combatants out of protective cover. The use of lethal chemicals then followed and escalated as systematic surveys indicated more potential agents and, as the WHO points out:[24]

> The chemical industry, not surprisingly, was a major source of possible agents, since most of the new chemical warfare agents had initially been identified in research on pesticides and pharmaceuticals.

So the critical link between civil research and military uses is again made quite clear.

The link to the pharmaceutical industry is important because, as drugs that could help people with some mental illnesses began to be discovered after the Second World War, the military became interested in the development of more incapacitating chemicals. As the US *Textbook of Military Medicine* commented:[25]

> Virtually every imaginable chemical technique for producing military incapacitation has been tried at some time.

and it went on to state that between 1953 and 1973, in the United States:[26]

> many of these were discussed and, when deemed feasible, systematically tested. Chemicals whose predominant effects were in the central nervous system were of primary interest and received the most intensive study.

The text suggests that virtually all drugs with prominent psychological or behavioural effects – psychochemicals – can be placed in four classes:

stimulants (for example, amphetamines), depressants (for example, barbiturates), psychedelics (for example, Lysergic acid diethylamide), and deliriants[27] that cause 'an incapacitating syndrome, involving confusion, hallucinosis, disorganized speech and behavior'. Amongst many such deliriants, chemical compounds that interfered with the ACh system – anticholinergics – were regarded as the most likely to be used as military incapacitating agents.

One of these anticholinergic deliriants, BZ (3-quinuclidinyl benzilate), was eventually weaponised by the United States and, as the text again makes clear, the process involved a transfer of civil research findings to the military:[28]

> BZ was first experimentally studied for therapy of gastrointestinal diseases. However, reports were received of confusion and hallucinations ... BZ was quickly withdrawn from commercial study and turned over to the U.S. Army as a drug of possible interest as an incapacitating agent.

BZ was produced between 1962 and 1965 and by 1970 there was a stockpile of 49 tons of the agent held by the United States. Although BZ had many shortcomings as an agent, and the stockpile was destroyed, eventually numerous other compounds with similar characteristics were investigated, for example by the United States and the United Kingdom.[29]

There are two main sub-classes of receptors for ACh in the nervous system, those affected by nicotine (nicotinic) and those affected by muscarine (muscarinic). BZ and similar agents latch on to the muscarinic type of receptor and thereby block the action of the ACh transmitter. As the US *Textbook of Military Medicine*[30] noted, 'The term anticholinergic ... refers more specifically to compounds that selectively block the brain's muscarinic receptor (now known to consist of several subtypes).'

More generally, the report of the Scientific Advisory Board (SAB) of the Organization for the Prohibition of Chemical Weapons (OPCW) to the Third Review Conference summarised the origins of candidate incapacitating chemicals as follows:[31]

> The types of chemicals and pharmaceuticals, known to have been considered as incapacitants, from open literature sources, were discussed. Most are centrally acting compounds that target specific neuronal pathways in the brain. All of them emerged from drug programmes undertaken from the 1960s to the 1980s, as far as can be judged by the research that has been published.

So the SAB, using only open sources, was able to show the strongest possible link between civil research and military developments in this field of neuroscience.

There is a clear consensus amongst many experts[32] that an operationally effective chemical incapacitant is not available at the present time, but there is also a concern that as our understanding of the chemistry of the brain – and how to manipulate it – develop some will believe that such an agent is possible and will therefore continue to seek to misuse civil research for hostile purposes.

Riot control agents act on the external sensory systems rather than the central nervous system, but there is also a concern that several states are seeking means for the long-range delivery of larger quantities of riot control agents[33] and of thereby producing weapon systems which might also be loaded with an incapacitating agent in situations other than riot control.

Biological weapons

Biological agents such as bacteria and viruses are capable of multiplying in the affected victim and thus differ crucially from chemical agents (and toxins, which are discussed later). So, many biological agents will be much more fragile in the environment than a chemical agent and thus more difficult to deliver effectively. On the other hand, only a very small amount may need to be delivered in order to cause an infection. Additionally, some biological agents can be contagious from the first victim to other people. Therefore, in addition to thinking about categories of lethal and non-lethal agents we have to think of contagious and non-contagious agents[34] (Table 1.2).

The link between civil research on infectious diseases and the development of biological weapons hardly needs stressing. The huge advances in our understanding of microbial pathogens, resulting from the work of people like Pasteur and Koch in the late nineteenth and early twentieth centuries, were made for beneficial purposes[35] but were

Table 1.2 Biological weapons agents

Potentially infectious from first victim	Incapacitating Lethal	e.g. Influenza virus e.g. *Yersinia pestis* (plague)
Not infectious from first victim	Incapacitating Lethal	e.g. *Coxiella burnetii* (Q-fever) e.g. *Bacillus anthracis* (anthrax)

then applied in numerous state-level offensive biological weapons programmes in the twentieth century. The hostile applications would obviously not have been possible without the civil work that characterised the bacteria in the first place. Similarly, the growing understanding of viruses was taken up in offensive programmes later in the twentieth century. Whilst the discussion here is focused on anti-personnel agents, it has to be said that the same argument can be made in regard to anti-animal and anti-plant biological warfare agents and offensive programmes.[36]

As we all know too well, when we are infected by a pathogen and become ill, our behaviour may change a good deal. As the WHO notes in regard to anthrax:[37]

> Inhalation anthrax begins with nondescript or influenza-like symptoms that may elude correct diagnosis. These may include fever, fatigue, chills, non-productive cough, vomiting, sweats, myalgia, dysponea, confusion, headache...followed after 1–3 days by the sudden development of cyanosis, shock, coma and death.

Given the intimate connections between the immune (defence) system and the nervous system such behavioural outcomes are to be expected. Similarly, tularæmia, caused by infection with *Francisella tularensis*,[38] usually results in 'an abrupt onset of fever, accompanied by chills, malaise and joint and muscle pain. Ulceroglandular tularæmia, caused by virulent strains, if untreated, has a case-fatality rate of about 5%'.

Other pathogens that have, like anthrax and tularæmia, been developed as biological warfare agents directly target the nervous system. Venezuelan equine encephalitis (VEE) virus is a member of the Alphavirus group and is transmitted naturally by mosquitoes but can also be infectious in a biological weapons aerosol. The related eastern equine encephalitis (EEE) and western equine encephalitis (WEE) viruses, which were also identified in the 1930s, cause more severe illnesses.[39] Considerable work with animals[40] has shown that in such models '[T]he lymphatic system and the CNS appear to be universal target organs...as was seen in humans'. As VEE is infectious in low doses by aerosol, can be produced at low cost and in large quantities, and is quite stable, it is not surprising that it was developed as an incapacitating agent during the twentieth century. So here again we can see the cycle of civil science advances being taken up and applied to other, hostile, purposes.

Toxins

It is necessary to begin here by noting that the understanding of the word 'toxin' in relation to the BTWC and CWC prohibition is different from that held by scientists. As the WHO explained:[41]

> In the sense of the Biological and Toxin Weapons Convention, 'toxin' includes substances to which scientists would not normally apply the term. For example, there are chemicals that occur naturally in the human body that would have toxic effects if administered in large enough quantity. Where a scientist might see a bioregulator, say, the treaty would see a poisonous substance produced by a living organism, in other words a toxin.

Histamine, for example, occurs naturally in the body, but its delivery in a sting, as we all know, is something different. Of course, since such a substance is a chemical, and toxic, it would automatically fall under the CWC prohibition.

There are, of course, many different types of toxin, which can be lethal or incapacitating. Examples of both kinds have been developed in offensive programmes as they specifically attack the nervous system. The bacterium *Clostridium botulinum* produces neurotoxins that have often caused food poisoning. The WHO stated:[42]

> Botulinum toxins are the most acutely lethal of all toxic natural substances. As a dry powder, they may be stable for long periods. They are active by inhalation as well as ingestion.

Not surprisingly, it then added, '[T]hey have long been studied as warfare agents of the lethal type, particularly, though not exclusively, types A and B'.

The US *Textbook of Military Medicine* explains the neurotoxic action of botulinum toxin as follows:[43]

> The extreme toxicity of the botulinum toxins would lead us to believe that it must have some highly potent and efficient mechanism of action. This probability made botulinum toxin the subject of work by many laboratories, especially after we learned that it is a neurotoxin. Experiments with *in vitro* neuromuscular models established that the toxin acts presynaptically to prevent the release of acetylcholine.

It also notes that one of the legacies of the military research on these toxins during the Second World War[44] was the development of 'the botulinum vaccine that is used even today'.

Staphylococcal enterotoxins produced by the bacterium *Staphylococcus aureus* are a very common cause of diarrhoeal food poisoning. The WHO stated of one such toxin that:[45]

> It is heat-stable and, in aqueous solution, can withstand boiling. It is active by inhalation, by which route it causes a clinical syndrome markedly different, and often more disabling, than that following ingestion. It has been studied as a warfare agent of the incapacitating type.

The US *Textbook of Military Medicine* describes the mechanism of action as follows:[46]

> When inhaled as a respiratory aerosol, SEB [staphylococcal enterotoxin B] causes fever, severe respiratory distress, headache, and sometimes nausea and vomiting. The mechanism of intoxication is thought to be from a massive release of cytokines.

SEB is, in fact, a superantigen that provokes this massive response from the immune system and thus indirectly affects the brain and behaviour.[47]

Toxins can be obtained from bacteria, marine organisms, fungi, plants and animal venoms. They come in many different types and sizes of molecule and attack diverse targets in the victim's body. Historically, bacterial toxins were the most important potential weapons agents because of their toxicity, but the US *Textbook of Military Medicine* commented that animal venoms 'must be considered potential future threats…as large-scale production of peptides becomes more efficient'.[48] This is of interest here, of course, because many such venoms attack the victim's nervous system in order to achieve rapid effects. However, the reference to peptide production is of most interest.

Concerns about novel peptide agents have been made clear by states party to the BTWC for over twenty years. The contribution of the United States to the background paper on relevant scientific and technological developments for the 1991 Third Review Conference stated, in part, that:[49]

> peptides are precursors of proteins made up of amino acids...They are active at very low concentrations...Their range of activity covers the entire living system, from mental processes (e.g. endorphins) to many aspects of health such as control of mood, consciousness, temperature control, sleep, or emotions, exerting regulatory effects on the body.

The text on these bioregulatory peptides, some of which clearly are involved in the operation of the nervous system, continues directly:

> Even a small imbalance in these natural substances could have serious consequences, including fear, fatigue, depression or incapacitation. These substances would be extremely difficult to detect but could cause serious consequences or even death if used improperly.

As if to reinforce the point, the Canadian Government took the unusual step of producing a separate document titled *Novel Toxins and Bioregulators: The Emerging Scientific and Technological Issues Relating to Verification and the Biological and Toxin Weapons Convention*[50] and sending it to all states party to the Convention. This reiterated the point made by the United States[51] that '[B]ecause bioregulators have many different sites of action, this gives rise to the possibility of selectively affecting mental processes and many aspects of health, such as control of mood, consciousness, temperature control, sleep or emotions'.

The Canadian study also noted that, given the evolutionary pressures for survival, toxins are likely to have reached an endpoint of selectivity for a particular target but:[52]

> with bioregulators, this is not the case since these compounds are involved in modulating cellular activities. They do not have a single endpoint of functions as neurotoxins do. The significance of this is that, while it is unlikely that research may lead to more toxic lethal agents, it may be possible to make more effective incapacitating agents.

The document goes on to discuss a wide range of toxins, and bioregulators such as Substance P, which subsequently were the subject of considerable discussion.[53,54,55] Indeed, the UK considered putting Substance P on the schedules of the CWC verification system to serve as a marker for such bioregulators, in the same way as saxitoxin and ricin serve as markers for toxins, and ensure that it is understood that all come under the General Purpose Criterion.[56]

In the last decades of the twentieth century the revolution in life and associated sciences continued apace and was applied in the large-scale, illegal, offensive biological weapons programme of the former Soviet Union. One focus of this programme was precisely on the hostile misuse of bioregulators. There is much that remains unknown about this programme, but a recent major study by Leitenberg and Zilinskas shows how dangerous such misuse could become. They argue that the 'most advanced and frightening research done in this area involved human myelin ... that acts as a type of insulator for nerves'.[57] The aim of the work was to engineer a bacterium so that it would produce a protein like myelin when it infected a human victim. The host's immune system would then mount an attack on myelin including that surrounding its own neurons to destroy the myelin. The authors explained[58] that '[W]ithout their myelin, nerve cells gradually lose their ability to send electrical signals. The result would be an artificial version of multiple sclerosis.' However, there would be a major difference from the natural disease:[59] '[The] difference between natural multiple sclerosis and the autoimmune disease induced by the genetically engineered *L. pneumophila* [bacterium] would be that the first takes years to kill its victims, whereas the second would progress to death in a matter of weeks.' Alibeck had given a brief account of this work previously,[60] but made the crucial point that '[A] new class of weapons had been found'. Clearly, this opened up numerous different possibilities for getting agents that produced foreign bioregulators into the victim in order to carry out hostile manipulation.

Conclusion

So it is quite clear that over the last hundred years the growth of civil life and associated sciences, including neuroscience, have been used in the development of chemical and biological weapons in ways that were not previously possible. Moreover, it seems highly probable that such transfers from civil to military applications will continue unless more effective means are developed to prevent that happening. With that history, it would be reasonable to expect that, even if they had ignored the possible misuse of their work before the 9/11 attacks and the subsequent sending of anthrax letters in the United States, neuroscientists would quickly have responded thereafter to ensure that they could engage in the process of preventing such misuse.[61] They could, for example, have ensured that biosecurity became part of the standard education given to all students of neuroscience. As the UK

Royal Society recommended in its 2012 study, *Neuroscience, conflict and security*:[62]

> Recommendation 1: There needs to be fresh effort by the appropriate professional bodies to inculcate the awareness of the dual-use challenge (i.e. knowledge and technologies used for beneficial purposes can also be misused for harmful purposes) among neuroscientists at an early stage of their training.

The available evidence strongly suggests that a great deal remains to be done to achieve that objective.[63]

References

1. Ferguson, N. (2006) *The War of the World: History's Age of Hatred.* London: Allen Lane.
2. Redmond, C. Pearce, M. J., Manchee, R. J. and Berdal, B. P. (1998) Deadly relic of the Great War. *Nature,* **393**, 747–748.
3. Dando, M. R. (1994) *Biological Warfare in the 21st Century.* Brassey's, London. (p. 65).
4. General Assembly (2013) *Resolution adopted by the General Assembly: 67/35 Measures to uphold the authority of the 1925 Geneva Protocol.* A/RES/67/35, United Nations, New York, 4 January.
5. Reference 3, p. 235.
6. Seventh Review Conference of the States Parties to the Convention on the Prohibition of the Development, Production and Stockpiling of Bacteriological (Biological) and Toxin Weapons and on their Destruction (2012) *Final Declaration.* BWC/CONF.VII/7, Geneva, 13 January.
7. Reference 3, p. 237.
8. For the text of the Convention see the website of the Organisation for the Prohibition of Chemical Weapons at <opcw.org>.
9. Conference of States Parties (2013) *Report of the Third Special Session of the Conference of the States Parties to Review the Operation of the Chemical Weapons Convention.* RC-3/3, OPCW, The Hague, 19 April.
10. Carlson, R. and Frankel, M. (2011) Reshaping responsible conduct of research education. *AAAS Professional Ethics Report,* **24** (1), 1–3.
11. Kelle, A., Nixdorff, K. and Dando, M. R. (2006) *Controlling Biochemical Weapons: Adapting Multilateral Arms Control for the 21st Century.* Basingstoke: Palgrave Macmillan. (pp. 116–137).
12. Williams, L. P. (1965) *Michael Faraday: A Biography.* London: Chapman and Hall. (pp. 107–108).
13. ibid, pp. 482–483.
14. World Health Organization (2004) *Public Health Response to Biological and Chemical Weapons: WHO Guidance,* 2nd Edition, Annex 1, pp. 143–213. WHO, Geneva.
15. Finger, S. (2000) *Minds Behind the Brain: A History of the Pioneers and their Discoveries,* chapter 16: Otto Loewi and Henry Dale: The Discovery of Neurotransmitters. Oxford: Oxford University Press.

16. Dando, M. R. (2006) *A New Form of Warfare: The Rise of Non-Lethal Weapons*. London: Brassey's. (pp. 69–70).
17. ibid, pp. 70–71.
18. Schmaltz, F. (2005) Neurosciences and research on chemical weapons of mass destruction in Nazi Germany. *Journal of the History of Neurosciences*, **15**, 186–209.
19. McLeish, C. and Balmer, B. (2012) Development of the V-series nerve agents, pp 273–288 in J. B. Tucker (Ed.), *Innovation, Dual Use and Security: Managing the Risks of Emerging Biological and Chemical Technologies*. Cambridge, MA: MIT Press.
20. ibid, p. 282.
21. Kristof, N. (1995) Hundreds in Japan hunt gas attackers after 8 die. *The New York Times*, 21 March, p. 1.
22. Sidell, F. R. *et al.* (1997) Vesicants. pp 197–228 in F. R. Sidell, E. T. Takafuji and D. R. Franz (Eds), *Textbook of Military Medicine*, Part I: Military Aspects of Chemical and Biological Warfare. Office of the Surgeon General, Department of the Army, Washington, D.C. (p. 212).
23. Reference 14, p. 144.
24. ibid, p. 143.
25. Ketchum, J. S. and Sidell, F. R. (1997) Incapacitating Agents. Pp 287–305 in F. R. Sidell, E. T. Takafuji and D. R. Franz (Eds), *Textbook of Military Medicine*, Part I: Military Aspects of Chemical and Biological Warfare. Office of the Surgeon General, Department of the Army, Washington, D.C. (p. 291).
26. ibid, p. 291.
27. ibid, pp. 292–294.
28. ibid, p. 295.
29. Dando, M. R. and Furmanski, M. (2006) Midspectrum incapacitant programs, pp. 236–251 in M. Wheelis, L. Rózsa and M. R. Dando (Eds), *Deadly Cultures: Biological Weapons Since 1945*. Cambridge, MA: Harvard University Press.
30. Reference 25, p. 294.
31. Conference of States Parties (2012) *Report of the Scientific Advisory Board on Developments in Science and Technology for the Third Special Session of the Conference of the States Parties to Review the Operation of the Chemical Weapons Convention*, RC-3/DG.1. OPCW, The Hague. (paragraph 12, p. 4).
32. Royal Society (2012) *Brain Waves Module 3: Neuroscience, conflict and security*. Royal Society, London.
33. Crowley, M. (2013) *Drawing the line: Regulation of 'wide area' riot control agent delivery mechanisms under the Chemical Weapons Convention*. Bradford Non-Lethal Weapons Project and Omega Research Foundation, University of Bradford, April. Available at <http://www.brad.ac.uk/acad/nlw/>.
34. Reference 3, p. 32.
35. Dando, M. R. (2006) *Bioterror and Biowarfare: A Beginner's Guide*. See chapter 2: Biological warfare before 1945, pp. 11–32, particularly table 2.1 on p. 16. Oxford: One World.
36. Reference 35, pp. 15–45.
37. Reference 14, p. 238.
38. ibid, p. 251.
39. Smith, J. F. *et al.* (1997) Viral Encephalitides. Pp 561–589 in F. R. Sidell, E. T. Takafuji and D. R. Franz (Eds), *Textbook of Military Medicine*, Part I: Military

Aspects of Chemical and Biological Warfare, p. 562. Office of the Surgeon General, Department of the Army, Washington, D.C.
40. ibid, p. 571.
41. Reference 14, p. 216.
42. ibid, p. 218.
43. Middlebrook, J. L. and Franz, D. R. (1997) Botulinum Toxins. Pp 643–654 in F. R. Sidell, E. T. Takafuji and D. R. Franz (Eds), *Textbook of Military Medicine*, Part I: Military Aspects of Chemical and Biological Warfare, p. 647. Office of the Surgeon General, Department of the Army, Washington, D.C.
44. ibid, p. 644.
45. Reference 14, p. 217.
46. Ulrich, R. G. *et al.* (1997) Staphylococcal Enterotoxin B and Related Pyrogenic Toxins. Pp 621–630 in F. R. Sidell, E. T. Takafuji and D. R. Franz (Eds), *Textbook of Military Medicine*, Part I: Military Aspects of Chemical and Biological Warfare. Office of the Surgeon General, Department of the Army, Washington, D.C. (p. 628).
47. Dando, M. R. (2001) *The New Biological Weapons: Threat, Proliferation and Control*. Boulder, CO: Lynne Rienner (pp. 61–2).
48. Franz, D. R. (1997) Defense Against Toxin Weapons, pp. 603–619 in F. R. Sidell, E. T. Takafuji and D. R. Franz (Eds), *Textbook of Military Medicine*, Part I: Military Aspects of Chemical and Biological Warfare. Office of the Surgeon General, Department of the Army, Washington, D.C. (p. 610).
49. Secretariat (1991) *Background Document on New Scientific and Technological Developments Relevant to the Convention on the Prohibition of the Development, Production and Stockpiling of Bacteriological (Biological) and Toxin Weapons and on their Destruction*, p. 29. BWC/CONF.III/4, United Nations, Geneva.
50. Canada (1991) *Novel Toxins and Bioregulators: The Emerging Scientific and Technological Issues Relating to Verification and the Biological and Toxin Weapons Convention*. Department of External Affairs and International Trade, Ottawa, September.
51. ibid, pp. 45–46.
52. ibid, p. 46.
53. Hamilton, M. G. (1998) Toxins: The Emerging Threat. *ASA Newsletter*, 98 (3), 20–26.
54. Koch, B. L. *et al.* (1999) Inhalation of Substance P and thiorphan: Acute toxicity and effects on respiration in conscious guinea pigs. *Journal of Applied Toxicity*, 19, 19–23.
55. Reference 47, chapter 4: Toxins, pp. 45–65 and chapter 5: Bioregulatory peptides, pp. 67–85.
56. Walker, J. R. (2012) *The Leitenberg-Zilinskas History of the Soviet Biological Weapons Programme*, pp 3–4. Harvard Sussex Program Occasional Paper No. 2. University of Sussex, UK, December.
57. Leitenberg, M. and Zilinskas, R. (2012) *The Soviet Biological Weapons Program: A History*. Harvard, MA: Harvard University Press (p. 194).
58. ibid, pp. 194–195.
59. ibid, p. 195.
60. Alibek, K. and Handleman, S. (1999) *Biohazard: The Chilling Story of the Largest Covert Biological Weapons Program in the World – Told from Inside by the Man Who Ran It*. New York: Random House. (p. 164).

61. Dando, M. R. (2009) Biologists napping while work militarized. *Nature*, **460**, 950–951.
62. See reference 32.
63. Walther, G. (2013) Ethics in Neuroscience Curricula: A Survey in Australia, Canada, Germany, the UK and the US. *Neuroethics*, **6** (2), 343–351.

2
The Structure and Function of the Brain

Introduction

This book is about critical questions of public policy, but in order to follow the argument it is necessary to have a basic outline of the growing understanding of the brain's structure and function.[1] People with diverse science and social science backgrounds are now entering neuroscience, or becoming interested in the security implications of neuroscience, so a brief introduction to some key points is provided here. A more detailed, but very accessible, introduction to the brain is available in *A Beginner's Guide to the Brain*.[2]

The brain is part of the human nervous system, which consists of the central nervous system (CNS), made up of the brain and spinal cord, and the peripheral nervous system (Figure 2.1). Input from peripheral sensory receptors, such as those in special organs like the eyes, is processed within the CNS after it is received via pathways travelling towards the CNS. These are the *afferent* pathways and are indicated on the left hand side of Figure 2.1. Output to muscles is then sent out from the CNS via what are called *efferent* pathways, shown on the right hand side of the figure. As Figure 2.1 shows, output goes to the voluntary skeletal muscles via the somatic nervous system, and to the visceral muscles and glands via the autonomic nervous system. Sensory receptors monitor what is happening in both these parts of the peripheral nervous system and feed this information back to the CNS.

The autonomic nervous system regulates body functions that are not normally under conscious control, such as heartbeat and digestion. A critical difference between this system and the somatic nervous system is that in the somatic system a motor (or action) neuron in the CNS can *directly* innervate a muscle, but in the autonomic system a message

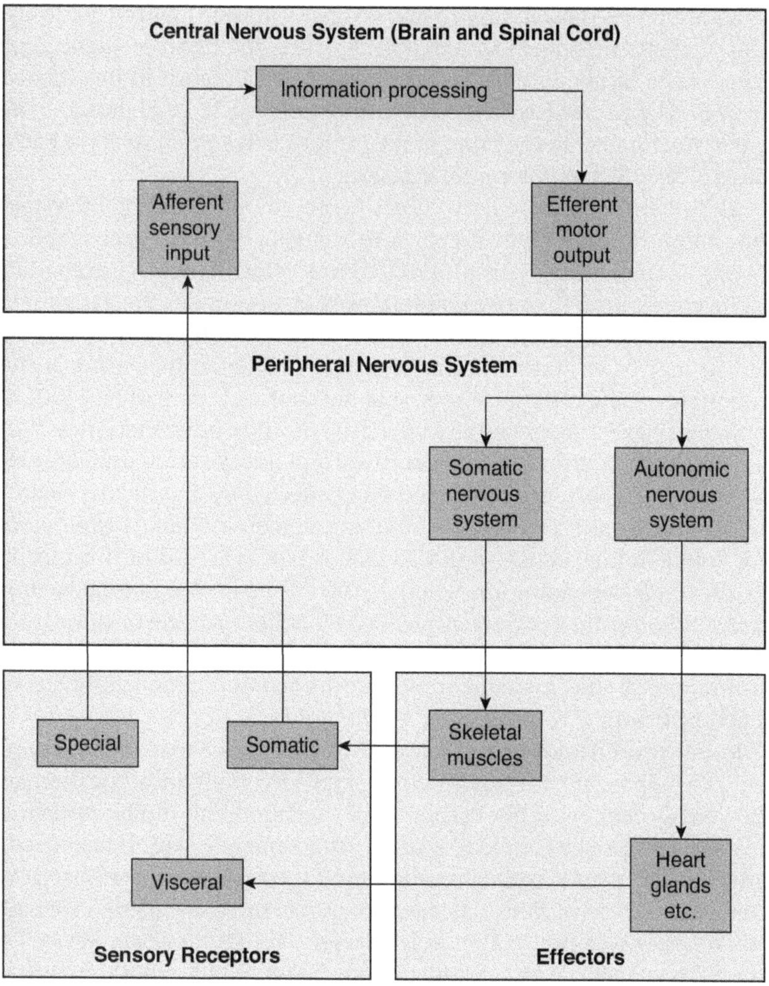

Figure 2.1 An overview of the functions of the nervous system

goes first to a peripheral autonomic ganglion (a collection of nerve cells) and then these cells in the peripheral ganglion directly innervate the relevant effector organ.

We can gain some idea of the role of neurotransmitters from a more detailed look at the autonomic nervous system. This system is divided into two parts called the sympathetic and the parasympathetic. These two parts act in opposition to each other so that in any particular situation if one is excitatory then the other will be inhibitory. Basically, the sympathetic branch is active in directing the fight, or flight, type response to a suddenly perceived threat. This will result in heightened awareness and reduced vegetative functions, such as digestion. The parasympathetic, on the other hand, is more active when we have had a large meal and are quiescently digesting it!

All the pre-ganglionic nerve fibres in the autonomic nervous system (i.e. those running from the CNS to the ganglia) have acetylcholine (ACh) as their neurotransmitter chemical, which is always excitatory in its effect. In the parasympathetic part of the system the ganglionic neurons themselves also have ACh as their neurotransmitter, but its effects may be excitatory or inhibitory depending on the nature of the receptors on the cells they innervate. By contrast, sympathetic ganglionic cells have noradrenaline (NA) (also known as norepinephrine, NE) as their neurotransmitter in most cases, and its effects are usually excitatory. The importance of the receptors affected by a particular neurotransmitter is easily seen in the autonomic nervous system. The receptors for acetylcholine are of two broad classes. These are called muscarinic or nicotinic depending on whether the effects of the natural neurotransmitter on them can be mimicked by either nicotine or muscarine (an extract from a fungus). The ACh parasympathetic synapses on the heart, for example, are muscarinic and the action of acetylcholine is to slow the heart.

In this short description of just part of the peripheral nervous system we can appreciate the enormous complexity built into the human nervous system, with the balancing of excitation and inhibition being carried out through different systems using different neurotransmitters, affecting different classes of receptor. And it has become ever clearer, as the genomics revolution has advanced, that there are many different sub-types of receptor in the broad classes. This kind of complexity, of course, is even more of a consideration when thinking about the central nervous system itself.

Our understanding of the complexity of the brain has recently grown through the discovery that two long-held views about it were, in fact, quite wrong. The first concerns the complexity of information transfer between neurons. It had been believed that whilst transmission of information *within* a neuron was by electrical means this was not generally the case as the transmission of information *between* neurons.[3] Communication between neurons was thought to be largely by chemical means – with a neurotransmitter released by the pre-synaptic neuron being recognised by receptors on the post-synaptic neuron. Modern work has shown that the situation can be much more complex, for example, groups of neurons can have tight electrical connections (called gap junctions) that serve to synchronise their activity, and the characteristics of some gap junctions can be modified chemically.

In line with this view of the complexity of information processing and integration in the human nervous system, it has also become clear that chemical neurotransmission between neurons is not always as simple as once thought. Rather than a neuron having only one type of neurotransmitter it now appears that there can be two transmitters and that one of these will often be a neuropeptide type (a peptide consists of a short string of amino acids) rather than a small molecule neurotransmitter like acetylcholine. It is also clear that some neurotransmitters can be produced, and act, outside of classical synapses to modify activity in other neurons. These kinds of finding suggest that it is now necessary to think of both neuromodulation and neurotransmission.

The second long-held view to be ousted was the idea that no new neurons are formed in the adult brain. For a number of years there had been evidence that new cells did arise in the brain, but this idea of neurogenesis was outside accepted dogma.[4] It took scientists time to demonstrate neurogenesis unequivocally, showing, for example, that male songbirds need the growth of new neurons to be able to sing in the spring and that, in addition to stress causing cell death in primates, release from stress and placement in rich environments can lead to neurogenesis. Indeed, it turns out that cells with stem cell-like properties are retained in the adult brain and this discovery has opened up all kinds of possibilities for new research and treatment. Neurogenesis is now one of the fastest growing areas of brain research.

Much of the human body is organised bilaterally, with paired structures like arms, hands, legs. The human CNS is no different and has paired structures such as the cerebral hemispheres, eyes and ears. The human brain is similar in structure to that of other primates, to which we are closely related, but has an expanded cerebral cortex overlying most of the older parts (in an evolutionary sense) of the brain. Readers will be familiar with the lateral view of the human brain shown in Figure 2.2, with the expansion of the brain lying at the top of the spinal cord. Other features of the brain may also be familiar, such as the primary motor area and the primary somatosensory area situated in the cerebral cortex, the cerebellum at the base of the brain and the gradual thickening of the spinal cord as it leads into the brain.

We can get an idea of how the CNS is organised by considering an imaginary slice taken down through the brain and spinal cord as in

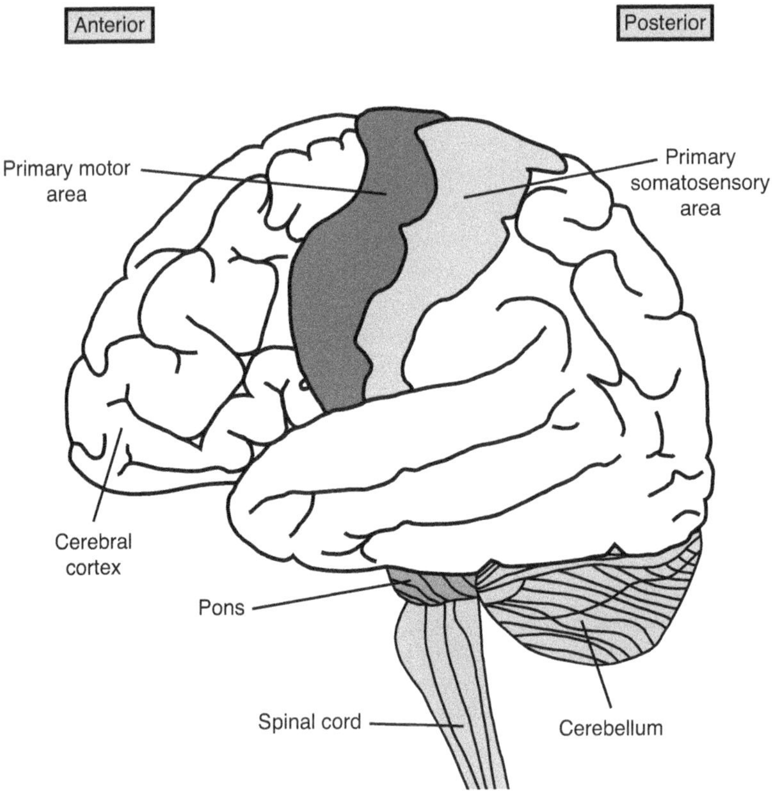

Figure 2.2 Simplified lateral view of the brain

Figure 2.3. Inside the CNS groups of neuron cell bodies are called *nuclei* and collections of nerve fibres running from such cell bodies are called nerve *tracts*. The route followed by information is called a *pathway*. We can follow such a pathway, starting from the first order sensory neuron in the lower left of Figure 2.3. This sensory neuron synapses with a second order sensory interneuron in the medulla oblongata at the top of the spinal cord. The fibre (axon) from this second neuron crosses over to the other side of the brain and terminates in the thalamus. The thalamus is the principle relay station for sensory input from the spinal cord to the cerebral cortex. Third order sensory interneurons convey the information to the primary sensory cortex. At the top left hand side of the figure a higher order motor neuron in the primary motor cortex is shown sending an axon down to the spinal cord. Again, this crosses over

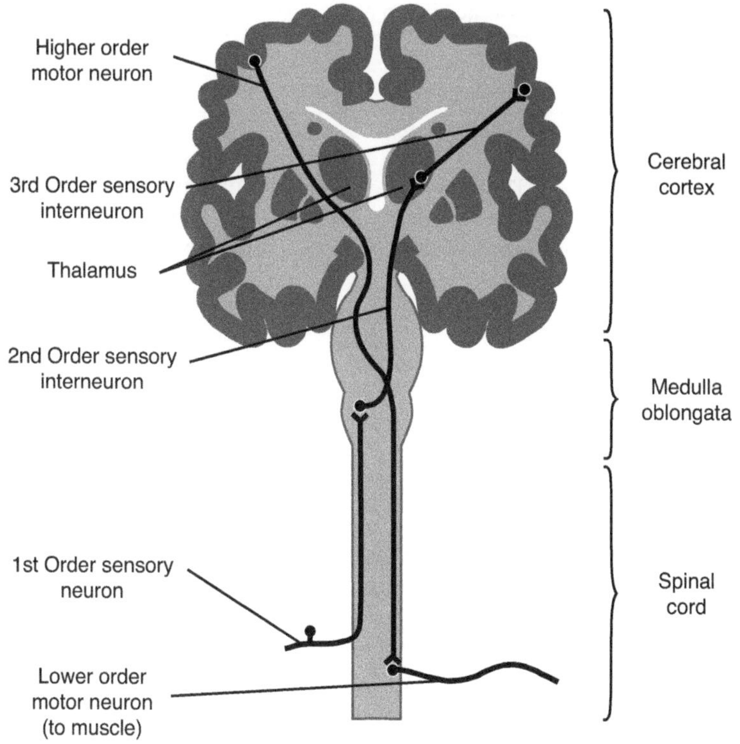

Figure 2.3 Simplified view of the sensory and direct motor pathways

in the basal part of the brain and synapses onto a motor neuron in the spinal cord which directly innervates the relevant muscle. In addition to this direct pathway, however, lower order motor neurons in the cord may receive input from, for example, the cerebellum, which also shapes the output to muscles.

The simplest way to understand the complex structure of the brain is by following its development around the brain vesicles (fluid filled spaces) in the embryo. These vesicles and some of the derived brain regions are shown in Table 2.1. As was seen in Figure 2.2, such a lateral view of the brain is dominated by the cerebral hemisphere and the cerebellum. These structures are overlain by a layer (cortex) of, so-called, *grey matter* which consists of nerve cell bodies. In the cerebral cortex these are concerned with complex functions such as language; in the cerebellum they are involved in the fine adjustment of motor output as indicated above.

Table 2.1 Development of the human brain

Primary brain vesicles	Secondary brain vesicles	Major derived brain regions
Prosencephalon (forebrain)	1. Telencephalon	1. Cerebral hemispheres 2. Basal ganglia
	2. Diencephalon	1. Thalamus 2. Hypothalamus
Mesencephalon (midbrain)	Mesencephalon	1. Tectum 2. Tegmentum
Rhombencephalon (hindbrain)	1. Metencephalon	1. Cerebellum 2. Pons
	2. Myelencephalon	Medulla oblongata

If these outer structures of cortex and cerebellum are stripped away in the side view, the underlying structure of different parts of the brain is revealed (Figure 2.4). The thalamus, as we have seen, is a critical centre between the spinal cord and the cerebral cortex that serves complex relay functions. Below the thalamus lies the hypothalamus which serves autonomic functions, such as hormone production and emotional regulation. The hypothalamus is intimately linked to the pituitary gland, which produces many important hormones. The midbrain has structures involved in, for example, the processing of visual and auditory information and the production of involuntary movements, such as the startle response to a loud noise. The hindbrain structures, such as the pons and medulla oblongata, contain neurons which control more autonomic functions, such as heartbeat and breathing rate.

The fine structure of the various parts of the brain is very complex. Some cerebral structures were described by early anatomists, such as the hippocampus, (called thus because its shape resembled a seahorse) and we shall later encounter other elements of the limbic system, such as the amygdala, and structures in the hindbrain, such as the locus coeruleus (LC). We do not need to go into great detail here, however, but only make a general point about the link between the nervous and hormonal systems. When we talk of the nervous system we are clearly dealing with an abstraction from reality. In an intact, functioning animal the operation of the nervous system has to be coherently integrated with the endocrine and immune systems. As the confidence of neuroscientists has increased, this more wide-ranging perspective has come to the fore, particularly in thinking about the control and impact of emotional factors.

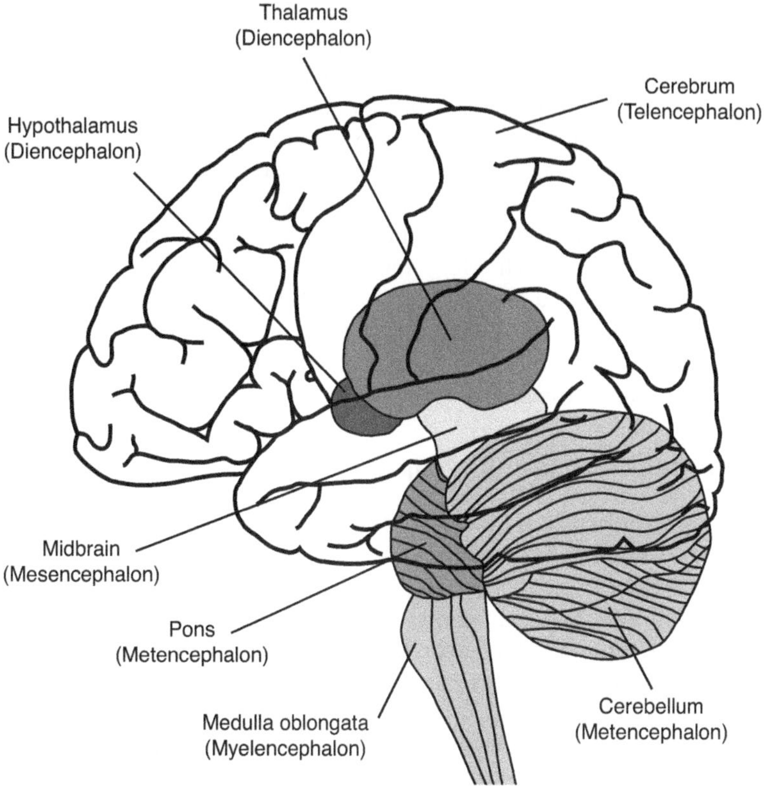

Figure 2.4 Some brain structures

Research over the last few decades has thus sought to answer questions such as 'Which systems underlie emotions?' and 'How does emotional processing in the brain interact with cognition, motor behaviour and motivation?'[5] The key system involved has become known as the limbic system. This comprises structures in the cortex and sub-cortical systems, such as the amygdala, that are strongly interconnected, often by reciprocal pathways.[6]

Once again, then, we are dealing with a complex system of interconnecting information flows. Given the complexity of the systems under study, and the range of data being acquired (from genetics to neuroimaging), it is hardly surprising that neuroscientists, having understood the importance of bioinformatics in the Human Genome Project, have been busy developing e-Neuroscience to add bioinformatics to their capabilities.[7] The Human Genome Project has generated huge amounts of new information on the sequence of the elements of DNA in the whole genome of our species and the genomes of many other species have been sequenced in the same way within just a couple of decades. Being able to store and analyse the vast amount of data – for example to search for similar genes in different species – would have been quite impossible without the powerful new computing capabilities developed in the information technology revolution. Neuroscientists have used these techniques in a like manner to search for DNA sequences similar to those for known G-protein-coupled receptors. It is very likely that this process will advance rapidly and considerably alter and improve our understanding of the brain.

Neurotransmitters and synapses

Much of the discussion in this book is concerned with specific transmitters like ACh and noradrenaline (NA) and how their operation in various neuronal circuits might be manipulated. Neurons come in all kinds of shapes and sizes. Figure 2.5 shows a typical example with short dendrites that receive input, the long axon that transmits the information and the axon terminals that release chemical neurotransmitters to pass on the information. The junction between two neurons at which the neurotransmitter is released is called a synapse, also shown in Figure 2.5. This illustrates the stored neurotransmitter on the pre-synaptic side of the junction and the post-synaptic membrane on which lie the receptor molecules for the neurotransmitter. The receptors can be fast acting (ion channel) or slower acting (G-protein) types. Clearing the transmitter from the synaptic junction may be done by means of reuptake into the pre-synaptic neuron or by enzymatic degradation in the synaptic cleft.

The usual description of a neuron and its interactions with other neurons goes as follows: the first neuron generates an impulse of electrical activity (called a nerve impulse or action potential) near the cell body (soma) and this travels down the long projection (axon) of the neuron until it reaches its ending where a neurotransmitter is released. This neurotransmitter affects receptors on projections from the second cell body (called dendrites) and, depending on the receptor and transmitter combination, either increases the possibility of initiation of an impulse of electrical activity (excitation) or reduces it (inhibition). The crucial point is that there is considered to be a rather specific relay of the message from neuron to neuron. In line with this idea of specificity,

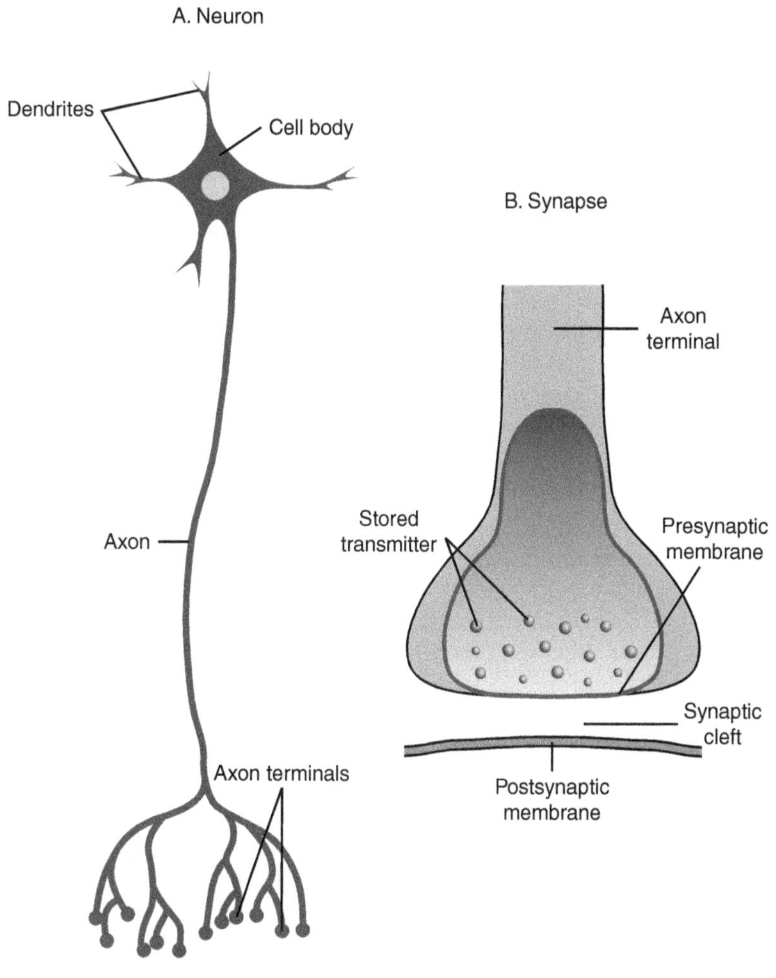

Figure 2.5 Diagrammatic representation of a neuron and a synapse

many tracts of similar axons (nerve fibres) in the brain have specific rather than widespread origins and destinations.

It has been discovered, however, that there is a very different system also present in the mammalian brain. In this second system relatively small numbers of nerve cells lie in groups in the lower parts of the brain and send their axon projections to ramify very widely in many other parts of the brain. This second type of organisation is now understood to have the function of modulating the activity of the more specific systems of the first type described above. The small cell groups of the lower brain have transmitters of six main types: norepinephrine (called noradrenaline NA here); dopamine; serotonin; epinephrine (or adrenaline); histamine and acetylcholine. In general, in relation to the modulatory system, the *Principles of Neural Science* notes that:[8]

> Each of these six neuronal systems has extensive connections in most areas of the brain and each plays a major role in modulating sensory, motor and arousal tone.

One of these modulatory systems utilises noradrenaline and much of this is produced in a group of cells located in the LC which has a major role in arousal. The system has been analysed in great detail in experimental animals such as the rat, but is also present in humans. The groups of NA cells are mostly located in the pons and medulla and are labelled in groups A1 to A7. For example, in the rat the A1 and A2 cells send their axons to the hypothalamus and are involved in the control of heart and hormonal functions. The A5 and A7 groups send their axons to the spinal cord and are involved in modulating reflexes and pain responses. However, most of the NA neurons are located in the locus coeruleus (A6 group) and these send their axons extremely widely to the cerebral cortex, cerebellum, brain stem and spinal cord. In humans where the brain contains many millions of neurons the bilateral LC contains only some 10,000 neurons on each side of the brain. However, these 20,000 neurons project their axons to almost every major region of the brain and are involved in the maintenance of vigilance and responsiveness to novel stimulation. The axons of these A6 cells from the locus coeruleus branch many, many times in the areas of the brain in which they terminate. So the pattern of innervation means that input from this NA system can have a synchronous modulatory effect.

The receptors for NA (and adrenaline) are called adrenoreceptors. There are two broad classes, termed alpha and beta adrenoreceptors, and sub-types within each class. Of particular interest for this discussion are the alpha 2 adrenoreceptors, which are inhibitory and may be located on the cell body of the neuron producing NA. Such receptors are therefore called autoreceptors. If such autoreceptors are present then the activity of the neuron will tend to inhibit itself. This appears to be what happens in the locus coeruleus:[9]

> alpha 2-autoreceptors are located on noradrenergic nerve terminals and on the cell bodies of noradrenergic neurons in the LC...These

> autoreceptors cause an inhibition of noradrenergic cell firing and a reduction in NE cell firing and a reduction in NE release from the terminals.

It is thus not difficult to see how an agonist drug would work. The text just quoted continues:

> This can be seen in the properties of dexmedetomidine...a recently introduced alpha 2-agonist with combined sedative, anxiolytic (antianxiety), and analgesic (pain-reducing) effects...The sedative and anxiolytic effects of dexmedetomidine are believed to be mediated by alpha 2-autoreceptors in the LC.

In short, the relatively selective alpha 2-agonist dexmedetomidine causes sedative effects by its actions on the neurons of the LC, inhibiting their activity and thereby reducing their excitatory input to the higher reaches of the brain.

That is the story in brief, but it must be understood that a much deeper understanding of the role of the LC is being developed which, in the future, may well allow much more specific interventions into its many mechanisms of operation. It is to this more detailed understanding that attention must now be turned.

Functions of the locus coeruleus system

Although the LC has been the subject of intensive international research over the last 40 years, knowledge of its operations is still incomplete, as are theories about its functions, and we should expect both to continue to evolve for some time yet. However, the enormous amount of research that has already been done has revealed much about the functions of the LC. One major recent review by Berridge and Waterhouse suggests that the LC-NA system serves two general functions:[10]

> First the system contributes to the induction and maintenance of forebrain neuronal and behavioral activity states appropriate for the acquisition of sensory information (e.g. waking).

Clearly this basic function is what the alpha 2-agonist would interrupt by inducing sedation. However, it appears that the system also has a more specific function. Berridge and Waterhouse continued as follows:

> Second, within the waking state, NE enhances and/or modulates the collection and processing of salient sensory information via actions on sensory, memory, attentional and motor processes.

Dysregulation of this second, more specific, set of functions could result in deficits seen in a variety of brain dysfunctions.

Despite the widespread ramifications of axons from the cells of the LC, detailed anatomical studies have shown that where it has been possible to trace the fibres – from about 1,500 neurons in each nucleus in the rat or the several thousands in the monkey – there is a clear organisation in that some areas of the brain are much more densely innervated than others. Additionally, it appears that there is organisation within the LC itself in that different parts of the cell group innervate different structures. The whole system, of course, gains enormous complexity because of the increasing number of different adrenergic receptor sub-types that are being discovered in the brain and which could be differentially affected by output of the neurotransmitter.

Input to the neurons of the LC comes from a variety of sources within the brain. Importantly, there is both anatomical and physiological evidence:[11]

> that the prefrontal cortex is a major source of afferent drive to the LC. The existence of this connection is particularly important insofar as it links circuits involved in higher cognitive and affective processes with the LC-efferent path.

There is much evidence that many neurotransmitters affect the LC neurons including, for example, acetylcholine, opiate peptides and hypocretin (orexin), of which we will have much more to say in the book. It is generally agreed that all LC neurons use NA as a transmitter but it has also been found that a number of neuropeptides, which could be functional as transmitters, are located in LC neurons. For example, many LC neurons contain both NA and the neuropeptide galanin. The variety of potential additional neurotransmitters again adds layers of complexity to the system.

Because the LC in rats and primates is homogeneous and compact its neurons have long been the subject of electrophysiological studies. It appears that they tend to be active in synchrony and that this is caused not by the release of chemical neurotransmitters but by electronic coupling of the cells through what are called gap junctions (tight connections between them). The result is that output from the LC can lead to a global release of NA throughout most of the brain.

LC neurons have two quite distinct modes of activity, they can fire regularly in a tonic mode or they can fire in bursts in a phasic mode. In the tonic mode the discharge rate is highest when the animal is active, lower when it is quiet, lower still during slow-wave sleep and almost silent during REM (rapid eye movement) sleep. The phasic mode is seen when the waking animal is overtly attending to a novel situation, but this response quickly habituates if the stimulus is repeated. Interestingly, activity of LC neurons is seen *before* changes in behaviour.

Vigilance is a measure of the sustained attention that is required for normal behaviours and survival. The relationship of the tonic and phasic outputs of LC neurons was studied in a vigilance task given to monkeys. The monkeys earned a reward when they detected and responded to a target stimulus hidden amongst a series of non-target stimuli. When they made a correct detection they received a food reward on pressing a lever. Phasic LC responses were preferentially evoked when the animal correctly responded to target stimuli. When the animal failed to respond to the correct stimulus phasic LC responses were not present. However, the phasic response and behaviour were dependent on the rate of tonic discharge at the time. At low tonic discharge rates the animals appeared drowsy and the target stimuli failed to elicit a phasic response; at high tonic LC discharge rates the animals exhibited increased eye movements and more frequent false judgements were made. Only at moderate tonic LC discharge rates was there optimal target detection and strong associated phasic responses.

Given that the LC neurons are involved in waking and attentional processes and have such widespread connections in many parts of the brain they have been thought by researchers to be implicated in a wide range of clinical problems. One such example was related to post-traumatic stress disorder (PTSD) but, given the complexity of the brain, it has so far been difficult to substantiate such links:[12]

> multiple cognitive and affective disorders have been posited to involve a dysregulation of noradrenergic neurotransmission.

But:

> In most cases, there is very little evidence indicating a direct, causal relationship between dysfunction of noradrenergic neurotransmission and a particular behavioral disorder.

So it might be concluded that whilst relatively crude manipulation of the LC-NA system with an incapacitant causing sedation is possible, the complexity of the human nervous system will preclude more sophisticated malign manipulation of such brain circuits. This conclusion would almost certainly be mistaken for it suggests that complex circuits cannot be understood by neuroscientists bringing the full weight of modern approaches to bear in the way Kandel and his colleagues did in regard to memory in the Nobel Prize-winning work on the simple sea slug.

Characteristics of neuronal circuits

Though the LC is present in humans much of the research work has been done on rodents whose brains, although having common features such as the LC system, are much simpler than ours. Even work on primates may not translate easily to human beings, so caution is always in order when making straightforward cross-species comparisons. Yet the search for fundamental characteristics of the nervous system may be greatly assisted, as we have seen, by the careful choice of a particular simpler system to study. In fact, there is a long history, and a probable extensive future, of the fruitful use of invertebrates (such as the sea slug) to study fundamental aspects of nervous systems.[13] Major breakthroughs in understanding, and the ability to manipulate complex activities in humans, can come about from research on well-chosen simpler systems.

This idea has certainly been put forward in regard to the operations of the LC by some researchers. In a paper entitled 'Network reset: a simplified overarching theory of locus coeruleus noradrenaline function' it was argued that there has been a tendency to produce ever more complex theories of how the LC works. There is a concept, long held by scientists, called Occam's Razor, which states that elaborate theories should not be constructed if a simpler theory will adequately explain the phenomenon under study. These researchers therefore advanced a theory of how the LC functions based on detailed work carried out on a much simpler invertebrate system. In their view, analogously to what had been found in the invertebrate system:[14]

> phasic activation of noradrenergic neurons of the locus coeruleus in time with cognitive shifts could provoke or facilitate dynamic reorganisation of target neural networks, permitting rapid behavioral adaptation to changing environmental imperatives.

In other words, there are circuits (networks of interlinked neurons) in the higher reaches of the brain which carry out different functions and the input of neuromodulatory NA can switch those neural networks in various ways to meet changing necessities. The researchers drew upon numerous studies, from the 1970s on, that have been made of the stomatogastric nervous system in crabs and lobsters and that work will need a brief examination to see why it was thought to be so relevant to the operations of the LC-NA system. First, however, it is as well to remember what neuroscientists are trying to achieve.

Neuroscientists wish to understand the brain as 'a neurochemical machine that enables us to think, act and perceive' and to understand that neurochemical machine they need to be able to integrate the results obtained from many different levels of analysis.[15] Having made enormous strides over the last decade or so in understanding the molecular and cellular levels and, increasingly, the neuroimaging of human brains, they wish to link these disparate levels of analysis together. To do that they need to focus attention on the circuits of neurons involved in various functions. These circuits are often termed *microcircuits*.

Work at the level of microcircuits – investigating the different functional modules of which the brain appears to be composed – has been slow because it is still technically difficult to do. The neurons involved have to be identified, a circuit diagram has to be determined and the way the neurons interact to produce behaviours has to be established at the level of the synapses between neurons. Finally, it is necessary to simulate the operations of the microcircuit in an artificial model system to check that the observed behaviour can really be accounted for properly by the model proposed.

Microcircuits come in different types but, remarkably, all those identified so far exhibit rhythmic behaviour. Amongst the easiest types of rhythmic behaviour networks of neurons to investigate are the, so-called, central pattern generator (CPG) networks which underlie repetitive behaviour, such as breathing, walking, swimming and the like.[16] The rhythmic output of nerve impulses from these CPGs is often generated by cells that have the intrinsic property of firing spontaneously in bursts of impulses. They are often also linked together electrically so that they fire in unision. So investigation of such systems in simpler organisms can reveal fundamental characteristics which may then be used to help elucidate more complex systems in the brain. What then can the study of such a simple CPG system tell us about the neuromodulation of microcircuits?

The stomatogastric nervous system of the Decapoda Crustacea (crabs and lobsters) consists of four interconnected ganglia spread around the foregut and thus somewhat isolated from the rest of the nervous system.

Output from these ganglia is generated in the form of rhythmic motor output to the muscles of the gut. These rhythms control the ingestion of food through the oesophagus, storage in the large 'cardiac sac', mastication of the food by the teeth of the gastric mill and filtration of the food by the pylorus as it passes to the midgut.

The single stomatogastic ganglion lies on the top of the foregut and has been very extensively studied.[17] It consists of only some 30 neurons which control the muscles driving the gastric mill and the pylorus. The pyloric rhythm is a pacemaker-driven central pattern in which three electrically coupled pacemaker neurons of the dilator phase fire together in rhythmic bursts of impulses to the muscles of the gut and simultaneously inhibit the other neurons in the pyloric network. Between these bursts of impulses when the pacemakers are not firing the other neurons in the circuit rebound from inhibition and are active until again inhibited by the bursting pacemaker neurons. The pyloric rhythm, as might be expected from the intrinsic bursting properties of the pacemaker neurons, is spontaneously active if the ganglion is isolated. It always displays a characteristic three-phased output as one of the non-pacemaker neurons recovers from pacemaker-induced inhibition faster than the rest and therefore fires before the rest. But the dominant feature of this CPG is the presence of the spontaneously and rhythmically active pacemaker neurons.

The motor output, which drives the gastric mill is much more variable than the pyloric rhythm and is often not present in an isolated preparation. This gastric mill output is more dependent on reciprocal inhibition amongst the cells of the network, rather than on pacemaker neurons. Both the pyloric and gastric rhythms can be affected by input from the numerous sensory systems that monitor movements of the gut and by input from higher centres in the animal's central nervous system. Also, a large number of neuromodulatory chemicals have been found in the stomatogastric nervous system of these animals.

From our viewpoint a crucial experiment on these stomatogastric neural networks was reported as early as 1991. It had already been shown that individual neurons could switch between different networks and that the pyloric and gastric mill networks could merge to form a single functional unit. In the 1991 publication, however, it was additionally shown that:[18]

> under an identified neuromodulatory stimulus, the CGP that produces swallowing-like behaviour of the foregut in lobsters is constructed *de novo* from neurons belonging to other CPGs.

The authors pointed out that:

> Consequently neurons operating independently as members of different circuits may be reconfigured into a new pattern-generating circuit that operates differently from the original circuits.

In this case, input from a single neuromodulatory neuron having diverse effects on various neurons in the same and different functional networks was found to be capable of producing an entirely different output. Furthermore, as the paper suggested, the change in output made biological sense as it was probably involved in moving food from the oesophagus into the cardiac sac (prior to crushing by the gastric mill). To achieve this both the pyloric and gastric mill neuronal networks were completely remodelled. It was also noted that the input not only had a rapid action on some parts of the neuron networks, but also longer-term consequences – some tens of seconds – before the separate, characteristic, pyloric and gastric mill rhythms were re-established.

It is not difficult to see why the authors of the 'Network reset' idea for the operational switching of behaviour by LC-NE output found this stomatogastric model system relevant. Much further work has been carried out on the neuromodulation of the stomatogastric system and other relatively simple invertebrate examples, and this work is producing an increasingly coherent understanding of neuromodulation in such systems.[19] It is clear that CPGs are subject to a great deal of neuromodulatory input and that this generates considerable flexibility of their output, including the kind of total remodelling just described. It is also clear that the effects of different modulators interact and that neuromodulatory neurons themselves can be subject to modulatory input from higher centres of the nervous system.

Given the rapid developments in our understanding of such model systems, it has to be asked how long it will take for the more complex circuits of the higher centres of the mammalian brain to be well understood? And a question that follows naturally is whether the malign manipulation of such circuits for hostile purposes can be prevented. We turn to that question in later chapters, but first we need to examine the development of the chemical and biological non-proliferation regime and the major challenges it faces today.

References

1. Dando, M. R. (1996) *A New Form of Warfare: The Rise of Non-Lethal Weapons*. London: Brassey's.

2. Al-Chalabi, A., Turner, M. R. and Delament, R. S. (2006) *A Beginner's Guide to the Brain*. Oxford: OneWorld Publications.
3. Bullock, T. H., Bennett, M. V. L., Johnston, D., Josephson, R., Marder, E., and Fields, R. D. (2005) The neuron doctrine, redux. *Science*, **310**, 791–793.
4. Miller, R. H. (2006) The promise of stem cells for neural repair. *Brain Research*, **1091**, 258–264.
5. Dalgleish, T. (2004) The emotional brain. *Nature Reviews: Neuroscience*, July, **6**, 582–589.
6. Morgane, P. J., Galler, J. R. and Molder, D. J. (2005) A review of systems and networks of the limbic forebrain/limbic midbrain. *Progress in Neurobiology*, **75**, 143–160.
7. Mortone, M. E., Gupta, A. and Ellisman, M. H. (2004) e-Neuroscience: challenges and triumphs in integrating distributed data from molecules to brain. *Nature Neuroscience*, **7** (5), 467–472.
8. Kandel, E. R., Schwartz, J. H. and Jessell, T. M. (2000). *Principles of Neural Science*, 4th Edition. New York: McGraw-Hill (see chapter 45, Saper, C.B., pp. 889–909, Brain Stem Modulation of Sensation, Movement and Consciousness).
9. Meyer, J. S. and Quenzer, L. F. (2005). *Psychopharmacology: Drugs, the Brain, and Behavior.* Sunderland, Massachusetts: Sinauer Associates (see chapter 5, pp. 119–138, Catecholamines).
10. Berridge, C. W. and Waterhouse, B. D. (2003). The locus coeruleus-noradrenergic system: modulation of behavioral state and state-dependent cognitive processes. *Brain Research Reviews*, **42**, 33–84 (p.34).
11. ibid, p. 37.
12. ibid, p. 64.
13. Sattelle, D. B. and Buckingham, S. D. (2006). Invertebrate studies and their ongoing contribution to neuroscience. *Invertebrate Neuroscience*, **6**, 1–3.
14. Bouret, S. and Sara, S. J. (2005). Network reset: a simplified overarching theory of locus coeruleus noradrenaline function. *Trends in Neuroscience*, **28** (11), 574–582.
15. Grillner, S., Koslov, A. and Kataleski, J. H. (2005). Integrative neuroscience: linking levels of analysis. *Current Opinion in Neurobiology*, **15**, 614–621.
16. Grillner, S., *et al.* (2005). Microcircuits in action: from CPGs to neocortex. *Trends in Neuroscience*, **28** (10), 525–533.
17. Marder, E. and Bucher, D. (2007). Understanding circuit dynamics using the stomatogastric nervous system of lobsters and crabs. *Annual Review of Physiology*, **69**, 13.1–13.26.
18. Megrand, P., Simmers, J. and Moulins, M. (1991). Construction of a pattern-generating circuit with neurons of different networks. *Nature*, **351**, 60–63.
19. Dickinson, P. S. (2006). Neuromodulation of central pattern generators in invertebrates and vertebrates. *Current Opinion in Neurobiology*, **16**, 604–614.

3
The CBW Non-Proliferation Regime

Introduction

As the sciences of chemistry and biology developed in the twentieth century the international community, as noted in the first chapter, agreed a series of arms control and disarmament treaties aimed at restricting the use of such weapons. However, they made a far from easy or rapid progress and, as the recent use of nerve agent in Syria demonstrates, is not yet successfully concluded.[1] Debate continues over why the US President called the use of such chemical weapons a 'red line' that should not be crossed. One reason is that such use opens up the possibility that these weapons, rather than being off limits, become an acceptable method of warfare. Then, all manner of novel weapons involving the misuse of the sciences of chemistry and biology could quickly be developed and used as the non-proliferation regime eroded.[2] The 1925 Geneva Protocol may be a very simple document by modern standards but, as shown in the first chapter, it does clearly state this principle of non-use and has become widely accepted as a customary international law binding on all states.

The Biological and Toxin Weapons Convention

Although after the Second World War the United Nations classified chemical and biological weapons with nuclear weapons as weapons of mass destruction it was not until the late 1960s, as the huge United States offensive biological and toxin weapons programme was drawing to a close, that the UK proposed that, rather than trying to deal with chemical and biological weapons together, biological weapons should be dealt with first. The UK draft convention proposed supplementing the Geneva Protocol with an article that not only banned the use of biological weapons but also banned research aimed at the production of such weapons. Neither of these elements survived into the agreed Convention, but after strong initial opposition to a separate treaty, the Biological and Toxin Weapons Convention (BTWC) was eventually agreed and entered into force in 1975.[3]

The BTWC, as described in Chapter 1, contains the crucial General Purpose Criterion in Article I that prohibits non-peaceful uses of biological agents. Despite being a short text of just fifteen articles (Table 3.1), it is a complex agreement with a number of interrelated facets. The premier historian of the BTWC, Nicholas Sims, has described these facets as the regimes of compliance, development and permanence and his analysis allows an insight into the function of the BTWC as a whole.[4]

The Convention is clearly a disarmament agreement. Articles I, II, and III do not allow the development, production, stockpiling or acquisition of agents or weapons, or the retention, or transfer, of such stocks. One of the two major weaknesses of the BTWC is that it has no system for verifying that States Parties are living up to these obligations, despite much of the 1990s being spent in trying to agree such a system.[5,6] Thereafter, in 2001–2002, 2006 and 2011 the Fifth, Sixth and Seventh Review Conferences of the BTWC mandated what have been called Intersessional Processes (ISPs), in which annual meetings address what appear to be more tractable issues.

Table 3.1 Summary of the Biological and Toxin Weapons Convention

Articles	
I.	Not to develop, produce, stockpile or acquire agents, weapons etc.
II.	To destroy stocks.
III.	Not to transfer to or assist others.
IV.	To take national measures.
V.	To consult and cooperate in solving problems.
VI.	May lodge complaint with the Security Council.
VII.	To provide assistance in the event of a violation.
VIII.	No detraction from the Geneva Protocol.
IX.	Obligation to continue negotiations on chemical weapons.
X.	Cooperation for peaceful purposes.
XI.	Amendment.
XII.	Review.
XIII.	Duration.
XIV.	Signature, ratification, deposition.
XV.	Languages.

In Sims' view, Articles IV, V, VI and VII provide functional substitutes for verification in this compliance regime. Whilst there has been little use or development of Articles V, VI and VII, it will become clear that Article IV has become of greater and greater importance, both because well-structured and performed national implementation of the Convention can provide considerable assurance to other States[7] and because, from the main point of interest here, effective in-depth implementation must involve the scientific community so that their expertise can be brought to bear on developing and maintaining the non-proliferation regime. Given its importance, it is hardly surprising that national implementation was selected as one of the three Standing Agenda Items (SAIs) to be addressed in the ISP Experts and States Parties Meetings in 2012, 2013, 2014 and 2015 in the lead up to the Eighth Review Conference of the BTWC in 2016.[8] Similarly, the annual data exchanges (voluntary confidence-building measures, CBMs) agreed in 1986 and 1991 at the Second and Third Review Conferences, despite their deficiencies in submission, continue to be a focus of potential development as a means of compliance assurance.

Article X of the BTWC has just two paragraphs that define what Sims terms the regime of development, and in his view:[9]

> There is much that can be criticized in the drafting of both paragraphs. They are so loosely constructed as to mean almost anything that a government or individual invoking them wants them to mean.

The first paragraph states, in part, that:

> The States Parties to this Convention undertake to facilitate, and have the right to participate in, the fullest possible exchange of equipment, materials and scientific and technological information for the use of bacteriological (biological) agents and toxins for peaceful purposes.

and the second paragraph states, in part, that:

> This Convention shall be implemented in a manner designed to avoid hampering the economic or technological development of States Parties to the Convention or international cooperation in the field of peaceful bacteriological (biological) activities, including the international exchange of bacteriological (biological) agents and equipment for the processing, use or production of bacteriological (biological) agents and toxins for peaceful purposes.

So, arguments about the relative weight of these paragraphs and Article III, in which States Parties undertake 'not to transfer' or 'assist, encourage, or induce' any other State or international organisation 'to manufacture or otherwise acquire any of the agents, toxins, weapons, equipment or means of delivery specified in Article I of the Convention', were almost inevitable. Clearly, developed States might view the BTWC as primarily a disarmament agreement, whilst developing States would be more interested in the economic benefits as the biotechnology industry became more important. Thus, a continuing theme in the evolution of the BTWC has been such disagreement, and Article X issues form a second element of the three Standing Agenda Items for the BTWC through to 2016.

The second grave weakness of the BTWC, in addition to its lack of an effective verification system, is that it has no large international organisation to take care of its maintenance between its five-yearly review conferences. Indeed, until 2006 it had no permanent organisation and since then its staffing has not even reached a total of ten people. How then can the BTWC be expected to remain in being in the choppy waters of international relations? It certainly states in the first paragraph of Article XIII that '[T]his Convention shall be of unlimited duration', but the article's second paragraph provides a mechanism for a State Party to withdraw if it wishes to do so – and with only three months notice.

Sims argues that perceptions are crucial: the preamble to the BTWC states that it is based on the determination to 'exclude completely the possibility' of biological and toxin agents being used as weapons. So the intentions of the negotiators are not in doubt.[10] He then goes on to stress the need to build on the 1925 Geneva Protocol by insisting that the small number of remaining reservations are removed by states and by doing all we can to strengthen the BTWC so that it is seen as embodying a norm of customary international law (and thus binding on even the small number of states who remain non-parties). However, the main means of development of the Convention has been through its five-yearly review conferences and the decisions taken and common understandings reached at those reviews.[11] Whilst these meetings have sometimes been very difficult, they have enabled states to keep the Convention developing in response to the ever-changing international situation.

It is in regard to these review conferences that Sims raises the question of a missing regime of research (as originally suggested by the UK). Sims points out that two understandings reached in the 1991 Third Review Conference clearly involve restrictions on research:[12]

> This Conference notes that experimentation involving open-air release of pathogens or toxins harmful to man, animals or plants that has no justification for prophylactic, protective or other peaceful purposes is inconsistent with the undertakings in Article I.

and:

> On the basis of the principle that sciences should support quality of life, the Conference appeals through State Parties to their scientific communities to continue to support activities that have justification under the Biological and Toxin Weapons Convention.

Indeed, Sims argues that a future review conference might declare that research and development are so closely linked that it is necessary that research is constrained in the same way as development. He is in no doubt about the difficulties that would arise, but suggests that this would at least shift the burden of proof to the researcher! The important point here is that consideration of the impact of scientific and technological change has become more and more important in meetings of States Parties to the BTWC. As the UK noted, in its contribution to the background paper on scientific and technological developments for the 2011 Seventh Review Conference, in a section on neuroscience:[13]

> Developments in this area could also result in the identification of compounds with potential for misuse as biological or toxin weapons since drugs acting on the brain to produce toxic or incapacitating effects could also have utility in a BW programme. Methods to facilitate delivery of such agents could also be exploited for harmful purposes, for example, to facilitate the entry of peptide neurotoxins across the BBB [blood-brain barrier].

After stressing that all such developments are covered by the scope of Article I of the Convention the UK continued:

> Since many of the benefits and risks of advances in neurosciences lie in the future, it is timely to consider issues related to governance of

> this dual-use technology area, balancing the obligations to take measures to prohibit misuse with the need to ensure that the beneficial development of science is not hampered.

The UK contribution noted that this was one reason why it had suggested that the annual ISP meetings before the Eighth Review Conference in 2016 should include more consideration of such scientific and technological issues and, as we shall see in later chapters, the review conference decided that the third SAI for these meetings would be on relevant developments in science and technology.

The Chemical Weapons Convention

Compared to the BTWC's five pages the Chemical Weapons Convention (CWC) is a massive document of several hundred pages.[14] In addition to its preamble and articles it contains several complex annexes (Table 3.2). Diplomats like to build on what they know well so, in addition to the presence of a General Purpose Criterion as discussed in Chapter 1, there are other obvious similarities between the CWC and the BTWC. For example, Article XVII deals with national implementation (like Article IV of the BTWC) and Article XI with international cooperation (like Article X of the BTWC). Yet there are, of course, major differences. Article VIII of the CWC, for example, deals with the Convention's international organisation – the Organization for the Prohibition of Chemical Weapons (OPCW) – and the Annex on Implementation and Verification deals with the verification system.

In his analysis of the BTWC at the start of the twenty-first century Nicholas Sims suggested that for the whole regime to develop and prosper there would be a need for the different regimes of compliance, development, permanence (and perhaps research) to be brought into a better balance. When the CWC came into force in the late 1990s the OPCW did not have a choice, it had to concentrate on one aspect of its role, disarmament.[15] Getting rid of biological and toxin agents is not difficult as they are unlikely to survive heating. Getting rid of chemical agents like nerve gases is difficult and very dangerous. Moreover, huge stocks of chemical weapons had built up in Russia and the United States during the East-West Cold War, and the destruction had to be organised and verified so that it was clear they had been destroyed.

For the purpose of verification, therefore, the chemicals of interest were placed into a series of schedules, and verification of what was done

Table 3.2 Summary of the Chemical Weapons Convention

Articles	
I.	General Obligations
II.	Definition and Criteria
III.	Declarations
IV.	Chemical Weapons
V.	Chemical Weapons Production Facilities
VI.	Activities not Prohibited Under this Convention
VII.	National Implementation Measures
VIII.	The Organization
IX.	Consultations, Cooperation and Fact-Finding
X.	Assistance and Protection Against Chemical Weapons
XI.	Economic and Technological Development
XII.	Measures to Redress a Situation and to Ensure Compliance, including Sanctions
XIII.	Relation to Other International Agreements
XIV.	Settlement of Disputes
XV.	Amendments
XVI.	Duration and Withdrawal
XVII.	Status of Annexes
XVIII.	Signature
XIX.	Ratification
XX.	Accession
XXI.	Entry into Force
XXII.	Reservations
XXIII.	Depositary
XXIV.	Authentic Texts
Annexes	
Annex on Chemicals	
Annex on Implementation and Verification	
Annex on Protection of Confidential Information	

with these chemicals was adjusted in relation to the dangers that they posed. So a Schedule I chemical, for example, would fit a set of criteria including:[16]

> (a) it has been developed, produced, stockpiled or used as a chemical weapon as defined in Article II;
>
> (c) it has little or no use for purposes not prohibited under this Convention.

So the chemicals listed under Schedule 1 included both well-known modern lethal agents and ricin and saxitoxin as markers for toxins (Table 3.3).

Table 3.3 Some CWC Schedule I Chemicals

Sarin
Soman
Tabun
VX
Mustard gas
Lewisites
Nitrogen mustards
Saxitoxin
Ricin

A Schedule 2 chemical would fit a list of criteria including:[17]

> (a) it poses a significant risk to the object and purpose of this Convention because it possesses such lethal or incapacitating toxicity as well as other properties that could enable it to be used as a chemical weapon;
>
> (c) it is not produced in large commercial quantities for purposes not prohibited under this Convention.

Thus BZ (3-quinuclidinyl benzilate) is included in this category.

A Schedule 3 chemical would fit criteria[18] that included, 'it has been produced, stockpiled or used as a chemical weapon' and it 'may be produced in large commercial quantities for purposes not prohibited under this Convention'. Thus phosgene and hydrogen cyanide were listed in this schedule.

Two points need to be clearly understood in regard to these schedules. Firstly, the General Purpose Criterion applies to *all* chemicals and thus the chemicals listed in the schedules are for the purposes of verification processes only. Secondly, exceptions are set out in Article II.9 of 'Purposes not prohibited under this Convention'. These include, for example, not only 'industrial, agricultural, research, medical, pharmaceutical or other peaceful purposes', but also:[19]

> (d) law enforcement including domestic riot control purposes.

This exemption has been the cause of much debate as some people have suggested that its natural reading shows that law enforcement is a larger category than domestic riot control and therefore there are law enforcement chemicals other than those standard riot control agents normally (and legally) used by police forces. Others have suggested that this reading would open up the possibility of novel incapacitating (so-called non-lethal) agents being developed, thereby leading to a progressive erosion of the whole Convention.[20] An example of the use of such chemicals was the deployment of fentanyl derivatives to end the 2002 Moscow theatre siege, so the concerns are not far-fetched.[21]

What is the threat?

Russian special forces used the fentanyl derivatives to deal with what was, of course, a very dangerous terrorist attack, but there has been increasing concern that terrorists could use chemical and biological weapons since the sarin attack on the Tokyo Underground and the sending of anthrax-contaminated letters in the United States following 9/11. The idea of a web of prevention – a set of policies that support the BTWC – had arisen earlier and included elements such as: good intelligence; export controls; effective national implementation of international agreements; sensible biodefence; and international determination to respond to deviation from the norm of non-use for hostile purposes.[22] However, as the perceived risks of terrorism have increased, other policies concerned with the safety and security of scientific work have begun to be introduced to prevent terrorist access to dangerous materials, technologies and knowledge.[23] Before we turn to that issue in the next chapter though, it is necessary to revisit the question of what threats all of these policies are intended to prevent.

Well-known calculations that are in the public domain show that, in the right circumstances, biological weapons could kill very large numbers of people or devastate agriculture. Chemical weapons may not be so devastating (as they do not have the ability to multiply in the victim or to be contagious from the first victim) but they are still seen as weapons of mass destruction.[24] It has recently become clear that chemical and biological weapons could be used on a smaller scale for terrorism or in the kinds of wars and interventions likely to characterise the coming decades of the twenty-first century.[25] In this book, the objective of the chemical and biological non-proliferation regime is understood to be the prevention of the use of such weapons for any of these purposes. But there is a crucial additional purpose, that of preventing the knowledge developed in the modern sciences of chemistry and biology from facilitating new generations of such weapons – able to be targeted at specific biochemical processes in humans, animals and plants – and thereby opening up a terrible new chapter in the history of human conflict.

This threat was spelled out by Professor Mathew Meselson of Harvard University in an article at the turn of the century. In his view:[26]

> During the century ahead, as our ability to modify fundamental life processes continues its rapid advance, we will be able not only to devise additional ways to destroy life but will also become able to manipulate it – including the processes of cognition, development, reproduction and inheritance.

And he continued:

> A world in which these capabilities are widely employed for hostile purposes would be a world in which the very nature of conflict had radically changed. Therein could lie unprecedented opportunities for violence, coercion, repression or subjugation.

Meselson also stressed that, unlike nuclear weapons, these capabilities cannot just be confined to states. In striking words he noted that 'biotechnology has the potential to place mass destructive capabilities in a multitude of hands', and 'to reach deeply into what we are and how we regard ourselves'. So he posed the question of whether we can prevent biotechnology going the way of *all* previous scientific and technological revolutions and being applied in a major way for hostile purposes. This book asks what neuroscientists, in particular, might do to prevent such a calamity.

References

1. Sample, I. and Borger, J. (2013) UK and France pass on evidence of sarin use in Syria. *The Guardian* (London), 5 June, p.1.
2. Perry Robinson, J. P. (2008) Difficulties facing the Chemical Weapons Convention. *International Affairs*, **84** (2), 223–239.
3. Dando, M. R. (1994) *Biological Warfare in the 21st Century: Biotechnology and the Proliferation of Biological Weapons*. London: Brassey's. (pp. 65–85).
4. Sims, N. A. (2001) *The Evolution of Biological Disarmament*. SIPRI Chemical and Biological Warfare Studies No. 19. Oxford: Oxford University Press.
5. Dando, M. R. (2002) *Preventing Biological Warfare: The Failure of American Leadership*. Basingstoke: Palgrave Macmillan.
6. Littlewood, J. (2005) *The Biological Weapons Convention: A Failed Revolution*. Aldershot: Ashgate
7. Canada, Czech Republic and Switzerland (2012) *National Implementation of the BTWC: Compliance Assessment: Update*. BWC/MSP/WP.6. United Nations, Geneva, 5 December.

8. Seventh Review Conference of the States Parties to the Convention on the Prohibition of the Development, Production and Stockpiling of Bacteriological (Biological) and Toxin Weapons and on Their Destruction (2012) *Final Declaration*. BWC/CONF.VII/7. United Nations, Geneva, 13 January.
9. Reference 4, p. 120.
10. Reference 4, pp. 151–170.
11. Pearson, G. S., Sims, N. A. and Dando, M. R. (2011) *Strengthening the Biological Weapons Convention: Key Points for the Seventh Review Conference*. University of Bradford, September.
12. Reference 4, p. 181.
13. Seventh Review Conference of the States Parties to the Convention on the Prohibition of the Development, Production and Stockpiling of Bacteriological (Biological) and Toxin Weapons and on Their Destruction (2011) *Background Paper: New Scientific and Technological Developments Relevant to the Convention*. BWC/CONF.VII/INF.3/Add.1. United Nations, Geneva, 23 November. (p. 31).
14. Organization for the Prohibition of Chemical Weapons (n.d) *Convention on the Prohibition of the Development, Production, Stockpiling and Use of Chemical Weapons and on Their Destruction*. OPCW, The Hague.
15. Kelle, A., Nixdorff, K. and Dando, M. R. (2012) *Preventing a Biochemical Arms Race*. Stanford: Stanford University Press.
16. Reference 14, p. 49.
17. ibid.
18. ibid, p. 50.
19. ibid, p. 5.
20. International Committee of the Red Cross (2012) *Toxic Chemicals as Weapons for Law Enforcement: A Threat to Life and International Law: i) Synthesis, ii) Summary*. Geneva: ICRC.
21. Riches, J. R. *et al.* (2012) Analysis of clothing and urine from Moscow theatre siege casualties reveals Carfentanil and Remifentanil use. *Journal of Analytical Toxicology*, **36**, 647–656.
22. Rappert, B. and McLeish, C. (Eds) (2007) *A Web of Prevention: Biological Weapons, Life Sciences and the Governance of Research*. London: Earthscan.
23. Kelle, A., Nixdorff, K. and Dando, M. R. (2006) *Controlling Biochemical Weapons: Adapting Multilateral Arms Control for the 21st Century*. Basingstoke: Palgrave Macmillan.
24. Reference 3, pp. 1–14.
25. See reference 2.
26. Meselson, M. (2000) Averting the hostile exploitation of biotechnology. *The Chemical and Biological Weapons Conventions Bulletin*, **48**, 16–19.

4
The Dual-Use Challenge

Introduction

In May 2013 the journal *Science* published a paper by Chinese scientists titled, 'H5N1 hybrid viruses bearing 2009/H1N1 virus genes transmit in Guinea pigs by respiratory droplet'.[1] The authors reported that:

> Using reverse genetics, we systematically created 127 reassortant viruses between a duck isolate of H5N1, specifically retaining its hemagglutinin (HA) gene throughout, and a highly transmissible, human-infective H1N1 virus.

So they concluded that 'hence, avian H5N1 subtype viruses do have the potential to acquire mammalian transmissibility by reassortment in current agricultural scenarios'.

Now these kinds of gain-of-function (GOF) experiments with deadly viruses have come under increasing criticism on scientific grounds. As was argued in *Nature* just one month before the Chinese authors published their paper,[2] today such viral combinations can certainly be made in an experimental system, but that does not mean that they will happen in the real world. A subsequent paper, in June 2013, with Anthony S. Fauci, Director of the US National Institute of Allergy and Infectious Diseases as one of the authors, put the point succinctly.[3] The authors noted that there are millions of exposures of humans to various avian influenza viruses daily, but as adaptations of such viruses to humans are extremely rare this suggests that:

> despite a low species barrier for infection, barriers against productive infection, and onward transmission, must be exceedingly high. The

> reason may be that to adapt fully to humans, avian influenza viruses require precisely attuned and mutually cooperative gene constellations... that are extremely unlikely to accumulate and survive in preadapted viruses.

In short, you can make an artificial virus in the laboratory, but in *nature* it is very unlikely indeed that such a virus would arise that could also overcome all of the host's defence mechanisms.

There are other scientific questions that might be asked about the quality of the work in the Chinese paper. Given that they were not using novel methods and the outcome of the research was practically inevitable, one might ask what was the point of doing it in the first place? However, that does not explain why Lord May of Oxford, a former UK Government Chief Scientist and past President of the UK Royal Society, was quoted as stating:[4]

> The record of containment in labs like this is not reassuring. They are taking it upon themselves to create human-to-human transmission of very dangerous viruses. *It's appallingly irresponsible.* [emphasis added]

As the paper in *Nature* had asked:[5]

> what if there were a leak or a small outbreak? Crippling lawsuits would follow. Are academic institutions sufficiently covered in terms of insurance? Are university regents or chancellors even aware of the power, and dangers, of the modern molecular biology going on in their labs?

The point being made, therefore, is that there could be *inadvertently caused* dangers in undertaking such experiments.

Such *biosafety* concerns have not, however, been the main focus of debate in regard to such experiments. Beginning in the early part of this century, after the 9/11 terrorist attacks and the sending of anthrax-contaminated letters in the United States, a series of experiments have raised concerns on the grounds of *biosecurity*. The life scientist undertaking, or publishing the results of, such experiments is facing a dilemma in that the materials, technologies and knowledge produced and published could do harm, not only inadvertently but also because of subsequent *deliberate hostile misuse*. Concern has been particularly evident in the United States as worries about large-scale bioterrorism have grown, but bioterrorism has also been the dominant framework in considerations of such *dual-use* experiments internationally.

A well-known example of media focus and scientific discussion was the mousepox experiment carried out in Australia by civil scientists seeking a new means of dealing with mouse plagues and the enormous consequent agricultural damage.[6] The scientists' idea was to modify the genome of a benign mousepox virus by inserting the gene for a mouse egg protein. The hope was that the infection would spread amongst the mice and production of the egg protein in female mice would provoke an immune response to their eggs. Consequently, the mouse plague would abate. The experiment worked, but not as well as the scientists had hoped, so they added a second gene to the modified mousepox – for the cytokine IL4 – in the hope that this would enhance the immune response. The result was a lethal virus that could overcome the defences even of a large percentage of mice vaccinated against the original mousepox. It did not take long, of course, for people to ask what might happen if the same genetic manipulation could produce a smallpox virus against which vaccination was ineffective?

Other experiments that caused concern and debate were the chemical synthesis of the polio virus and the reconstruction of the Spanish influenza virus.[7] However, the reason that the Chinese experiment of 2013 caused such a strong reaction was the debacle amongst the science community caused the previous year when it became clear that work financed by the US National Institutes of Health had been carried out in order to make deadly H5N1 influenza transmissible through the air between mammals.[8] This could hardly be dismissed as irrelevant from a biosecurity point of view since a highly pathogenic contagious influenza can be an extremely grave threat. Nevertheless, this did not appear to have been taken into account in either the funding or execution of the work and only became an issue when papers by groups in the Netherlands and the United States were submitted for publication in the journals *Science* and *Nature*.

The papers were sent for review to the US National Science Advisory Board for Biosecurity (NSABB), set up some years previously, which raised considerable concerns. A very public squabble then broke out over whether the papers should be published, and in what form. Eventually, both papers *were* published, but the paper by the Netherlands group only after a majority vote of 12 to 6 by the NSABB.

The dominance of the question of potential dual-use implications of single experiments in the debate on biosecurity in the life sciences community is often traced back to two important reports by the US National Academies in 2004 and 2006. However, even a short account of these two studies reveals that they have a much more wide-ranging and complex insight into the problem of biosecurity as the revolution in the life sciences continues in coming decades.

The first report was produced by a committee chaired by Gerald Fink (hence the Fink Report) and titled *Biotechnology Research in an Age of Terrorism*.[9] The committee clearly defined dual-use in the first page of its executive summary, stating that 'biotechnology represents a "dual-use" dilemma in which the same technologies can be used legitimately for human betterment and misused for bioterrorism'. But they went on to point out that the misuse of biotechnology could take many forms and that their report was *only* concerned with 'the capacity for advanced biological research activities to cause disruption or harm, potentially on a catastrophic scale'. So they were not concerned with other possible misuses of biotechnology, for example with the uses of older technologies that had underpinned the huge, offensive, biological weapons programmes of states in the last century.

Moreover, whilst the focus of the report was on the regulation of potentially dangerous experiments in the United States, the *international context* was made abundantly clear in the second page of the executive summary:

> Without international consensus and consistent guidelines for overseeing research in advanced biotechnology, limitations on certain types of research in the United States would only impede the progress of biomedical research here.

These two careful restrictions on the concept of dual-use – as a means of helping to minimise the potential misuse of the life sciences for hostile purposes – are often forgotten.

The Fink Committee is best known for its fourth recommendation that led to the setting up of the NSABB and for its second recommendation that seven whole classes of microbiological experiments were of sufficient dual-use concern as to warrant review on the grounds of biosecurity. These experiments included those that 'Would alter the host range of a pathogen ... Altering the tropism of viruses would fit into this class.' So they would obviously include the gain-of-function experiments with which we began this chapter.

It is particularly important, however, to note that the committee's first recommendation read as follows:

> **Recommendation 1: Educating the scientific community**
>
> We recommend that national and international societies and related organizations and institutions create programs to educate scientists about the nature of the dual-use dilemma in biotechnology and their responsibilities to mitigate its risks.

This recommendation presumably follows directly from the committee's realisation that the system of oversight they wished to see put in place could not work unless scientists understood the dangers, and their realisation that such understanding was largely absent within the community.

The second National Academies report was produced by a committee chaired by Stanley Lemon and David Relman (the Lemon-Relman report) and titled *Globalization, Biosecurity, and the Future of the Life Sciences*.[10] This committee built on the work of the Fink Committee and the idea of dual-use but, as the title of the report makes clear, it was particularly concerned with the impact of the worldwide spread of capabilities and the implications of the future trajectory of the life sciences. What is important from the perspective of this book on neuroscience is the Committee's second recommendation that:

> The committee recommends adopting a broader perspective on the 'threat spectrum'...so as to include, for example, approaches for disrupting host homeostatic and defense systems.

As the committee noted in its executive summary, 'the immune, neurological and endocrine systems are particularly vulnerable to disruption by manipulation of bioregulators', and it went on to point out that:

> bioregulators, which are small, biologically active compounds, pose an increasingly apparent dual-use risk. This risk is magnified by improvements in targeted delivery technologies that have made the potential dissemination of these compounds much more feasible than in the past.

So, in the view of the Lemon-Relman report, advances in *neuroscience* could be of equal concern to those in *microbiology* and *synthetic biology* (as in the mousepox experiment and the gain-of-function research on deadly influenza viruses).

It might be argued that since the Fink and Lemon-Relman reports were written and published some ten years before the Chinese experiments, they have had little real impact. Yet the original Fink Committee report did not envisage anything other than incremental change, even within the US. It suggested that its recommendations should be translated into an initial set of oversight guidelines and then[11] that these should be improved and updated 'as needed as…experience with the process grows'.

Experience grew slowly but the debacle over the transmissible H5N1 influenza experiments in 2012 accelerated the process, with Congress becoming increasingly involved,[12] and the US Government issuing a draft set of guidelines for oversight of dual-use research of concern (DURC). Such research is defined as:[13]

> life sciences research that, based on current understanding, can be reasonably anticipated to provide knowledge, information, products, or technologies that could be directly misapplied to pose a significant threat with broad potential consequences to public health and safety, agricultural crops and other plants, animals, the environment, materiel, or national security.

Although it will not happen overnight, it seems likely that as more experiments of dual-use concern come to public notice, oversight systems will be put in place in different countries, and that if scientists do not become proactively involved such systems will be implemented without giving much consideration to the views of scientists on the value of free and open research and publication. Moreover, neuroscientists should not imagine that they will be immune from this process. For example, the US draft regulations have work with botulinum neurotoxin designated as being within its scope and state that,[14] 'for the purposes of this Policy, there are no exempt quantities of toxin. Research involving any quantity of Botulinum neurotoxin should be evaluated for DURC potential.'

In the future it may seem that the raising of the issue of dual-use has generally had a positive effect in alerting the life sciences and policy communities to the need for them to cooperate in devising ways to help minimise the possibilities of hostile misuse. Yet it is important to be aware of the limitations of this approach. Much of the discussion of DURC has been in relation to synthetic biology but, as Jonathan Tucker cogently argued, there is a good deal of tacit knowledge required for success in that field and, even if a deadly virus could be produced, there are likely to be considerable problems in its effective weaponisation.[15]

In any concept of a problem, just as important as what it embraces and how it deals with that, is what it leaves out. It can be argued that, since the Fink and Lemon-Relman reports, the idea of dual-use has narrowed down rather than kept to their broader formulation. As one recent review noted, the dual-use issue is both broader and deeper than a simple dilemma about a single experiment. Rather, as the authors put it:[16]

> the research process has multiple stages from research design and funding application through to publication. The dual-use issue is relevant to all actors who impact on these stages.

And, perhaps more importantly:

> the issue is deeper, as it involves institutional norms and practices, including national and international models of scientific governance and funding.

It is therefore wrong just to focus on the individual scientist and a single experiment. The scientist certainly has a responsibility to be aware of the possibility of dual-use, but it may be far more important for scientists to engage their expertise in helping to fashion sensible policies at many different levels.

An example of such engagement can be seen in the request made by scientists in the Foundation for Vaccine Research (FVR) for the US Presidential Commission for the Study of Bioethical Issues to rule on whether making viruses more deadly than those found in nature is 'morally and ethically wrong'. Thus, these scientists have raised the issue of whether the critical questions about what is being done by the whole scientific, institutional, governmental and intergovernmental community in regard to the future of the life sciences has been lost in the debate on dual-use.[17] It seems likely, as events unfold in the manner Meselson foresaw (see Chapter 3), that questions regarding ethics and morality must be addressed more and more urgently.

Any idea neuroscientists might have that this is of little direct concern to them was dispelled by two papers published in late 2013 on the discovery of a new type of botulinum neurotoxin.[18,19] The two papers were accompanied by three editorial commentaries. One of these, by Michel Popoff of the Institut Pasteur in Paris, explains the historical context and scientific importance of this discovery.[20] However, it is the other two commentaries that are of particular interest here.

The first, with the title 'Novel *Clostridium botulinum* toxin and dual-use research of concern issues', explains that the journal's usual practice is to insist that:[21]

> It is general JID [*Journal of Infectious Diseases*] policy that gene nucleotide sequences must be submitted to the International Nucleotide Sequence Databases and that the accession numbers must appear in the final revision and published version of the manuscript.

In this case, nevertheless, after consultation with US government representatives from many different agencies and the JID editors, the authors were allowed to publish their papers 'while withholding the key gene sequences until appropriate countermeasures were developed'.

The second of the two additional commentaries is by David Relman who had co-chaired the US National Academies report, *Globalization, Biosecurity and the Future of the Life Sciences*.[22] He explained that there was good reason to withhold the information[23] as 'this new toxin, BoNT/H, cannot be neutralized by any of the currently available antibotulinum antisera'. This means, of course, that until a new specific antiserum is developed, misuse could pose a serious threat to public health. Relman agrees with the decision to withhold this information temporarily until countermeasures are in place but also argues that this is a rather simple case. Indeed, the core of his commentary is a consideration of the much more difficult cases that are likely to arise in the future as advances in the life sciences continue apace.

Relman begins by recalling the influential 1980s Corson report, *Scientific Communication and National Security*, that is principally remembered for its argument that there is a 'bright line' between scientific information that should be classified on security grounds and everything else that should be published openly to advance the progress of science. This viewpoint from the Cold War period, when nuclear issues were the dominant concern, influenced policy in the United States and elsewhere over the following decades.

Relman points out that what is often forgotten is that Corson also argued that there could be a grey area where limited restrictions short of classification were appropriate. Such research would have military applications that could be quickly applied and would provide a military advantage to opponents and would not be known to such adversaries already. Corson's committee recommended an alternative mechanism of voluntary publication control to deal with such research. Relman noted that such grey area research was ignored in subsequent years:[24]

> because there were few concrete and compelling examples of work that might fit in this category

but also:

> because the practical aspects of a nonclassification information control mechanism were, and remain, profoundly challenging.

The first of these reasons, as Relman points out, no longer applies and will apply less and less in coming decades. The second reason remains a profound challenge to the scientific and security communities.

Finding a way to deal with these issues will not be easy and, as Relman ends his commentary by stating,[25] for voluntary controls to be part of the eventual solution 'scientists will first need to recognize their ethical and moral responsibilities to society in the pursuit of knowledge'. He adds:

> Scientists have obligations to society that involve more than blind pursuit of knowledge. Like clinicians, *scientists have an obligation to do no harm*. [emphasis added]

We will return to the question of how voluntary control might fit into an overall system of minimising the potential misuse of benignly intended research in Part III of this book. In Part II we turn to the question of how modern advances in neuroscience might be misused.

References

1. Zhong, Y. *et al.* (2013) H5N1 Hybrid viruses bearing 2009/H5N1 virus genes transmit in Guinea pigs by respiratory droplet. *Science*, **340**, 1459–1463 (p. 1459).
2. Wain-Hobson, S. (2013) H5N1 viral-engineering dangers will not go away. *Nature*, **495**, 411.
3. Morens, D. M., Taubenberger, J. K. and Fauci, A. S. (2013) Pandemic influenza viruses – hoping for the road not taken. *The New England Journal of Medicine*, **368**, (25), 2345–2348.
4. Connor, S. (2013) 'Appalling irresponsibility': Senior scientists attack Chinese researchers for creating new strains of influenza virus in veterinary laboratory. *The Independent*, 1, 2 May.
5. Reference 2, p. 411.
6. Shinomiya, N., Minehata, M. and Dando, M. R. (2013) Bioweapons and dual-use research of concern. *Journal of Disaster Research*, **8** (4), 654–666.
7. ibid, pp. 659–661.

8. Novossiolova, T. A., Minehata, M. and Dando, M. R. (2012) The creation of contagious H5N1 influenza virus: implications for the education of life scientists. *Journal of Terrorism Research*, **3** (1), 39–51.
9. National Academies (2004) *Biotechnology Research in an Age of Terrorism*. Washington, D.C: National Academies Press.
10. National Academies (2006) *Globalization, Biosecurity, and the Future of the Life Sciences*. Washington, D.C: National Academies Press.
11. Reference 9, p. 7.
12. Grotton, F. and Shea, D. A. (2012) *Publishing Scientific Papers with Potential Security Risks: Issues for Congress*. R42606, Congressional Research Service, Washington, D.C., 12 July.
13. United States (2013) *United States Government Policy for Institutional Oversight of Life Sciences Dual Use Research of Concern*. Office of the Director, National Institutes of Health, Washington, D.C.
14. ibid, p. 7.
15. Tucker, J. B. (2011) Could terrorists exploit synthetic biology? *The New Atlantis*, **31**, 69–81.
16. Edwards, B., Revill, J. and Bezuidenhout, L. (2013) From cases to capabilities? A critical reflection on the role of 'ethical dilemmas' in the development of dual-use governance. *Science and Engineering Ethics*, **20** (2), 571–582.
17. Ross, R. (2013) Scientists seek ethics review of H5N1 gain-of-function research. *CIDRAP News*, 29 March.
18. Barash, J. R. and Arnon, S. S. (2013) A novel strain of *Clostridium botulinum* that produces Type B and Type H botulinum toxins. *Journal of Infectious Diseases*, **209**, (2), 183–191.
19. Dover, N., Barash, J. R., Hill, K. K., Xie, G. and Aron, S. S. (2013) Molecular characterization of a novel botulinum neurotoxin Type H gene. *Journal of Infectious Diseases*, **209**, (2) 192–202.
20. Popoff, M. R. (2013) Botulinum neurotoxins: More and more diverse and fascinating toxic proteins. *Journal of Infectious Diseases*, **209**, (2), 168–169.
21. Hooper, D. C. and Hirsch, M. S. (2013) Novel *Clostridium botulinum* toxin and dual use research of concern issues. *Journal of Infectious Diseases*, **209**, (2), 167.
22. See reference 10.
23. Relman, D. (2013) "Inconvenient truths" in the pursuit of scientific knowledge and public health. *Journal of Infectious Diseases*, **209**, (2), 170–172.
24. ibid, p. 171.
25. ibid, p. 172.

Part II
The Present

5
Modern Civil Neuroscience

Introduction

There is no doubt that the general public believes that advances are being made in neuroscience and in our understanding of how the brain works. There are numerous reports in the media about the work of civil neuroscientists. As an example, Table 5.1 lists some major articles from the popular science journal, *New Scientist*, in August and September of 2013, when sarin was being used on a large scale in Syria.[1]

What is the relationship between this media attention – particularly in the popular press – and what is actually being done and achieved by neuroscientists? One way of trying to answer that question is, as ever, to follow the money. Specifically, are there major new initiatives being made to fund neuroscience research on the scale of the Human Genome Project and what are the objectives of this funding? In 2013 two such funding programmes were announced, one in the United States and one in the European Union. A review of these programmes gives us an insight into what is being attempted by modern neuroscience.

Table 5.1 Examples of articles on neuroscience from *New Scientist* in August/September 2013

1. Birkhead, T. (2013) *Bird senses: Instant expert.* 3 August, pp. i–viii.
2. Heaven, D. (2013) *Higher state of mind: we have created a completely new form of intelligence...although no human can understand it.* 10 August, pp. 33–35.
3. Ananthaswamy, A. (2013) *Like clockwork: Our brains may run mechanically, like the springs and cogs in a finely tuned watch.* 31 August, pp. 33–35.
4. Bayne, T. (2013) *Special Issue. Thought: What are thoughts made of, and other mind-blowing questions.* 21 September, pp. 32–39.

The US BRAIN Initiative

In February 2013 *The New York Times* carried an article titled 'Obama seeking to boost study of the human brain'. The article reported that the aim of the project was to investigate the workings of the brain and thereby build a comprehensive map of its activities,[2] in short, to do what the Human Genome Project did for genetics. The scientists involved said that they hoped that federal funding would amount to $3 billion over a decade. There was clearly an economic motive for the investment as the article pointed out that the $3.8 billion invested in the Human Genome Project had netted the United States some $800 billion by 2010.

The BRAIN Initiative (Brain Research through Advancing Innovative Neurotechnologies) was announced in April 2013 with a proposed initial funding of $100 million in the fiscal year 2014. In order to ensure a swift start, the Director of the National Institutes of Health (NIH) convened an advisory group to identify high priority research areas that could be considered for this funding and the group produced an interim report in September 2013.[3] We can therefore use this report to identify what eminent neuroscientists believe is important, and feasible, research at the present time and whether, in part, they will consider potential dangers for hostile misuse arising from the research.

The interim report identifies nine research areas as having high priority for funding in 2014. These are shown in Table 5.2.

These priority research areas clearly range across the research spectrum from molecular to whole animal behaviour analysis, but there is also a major focus on intermediate levels of analysis – the neuronal circuits underlying behaviour and how they might be manipulated for investigational and medical purposes.[4] The executive summary of this

Table 5.2 High priority areas identified in the interim report

1. Generate a census of cell types
2. Create structural maps of the brain
3. Develop new large-scale network recording capabilities
4. Develop a suite of tools for circuit manipulation
5. Link neuronal activity to behavior
6. Integrate theory, modeling, statistics, and computation with experimentation
7. Delineate mechanisms underlying human imaging technologies
8. Create mechanisms to enable collection of human data
9. Disseminate knowledge and training

Source: Modified from Advisory Committee to the NIH Director (2013) *Interim Report: Brain Research through Advancing Innovative Neurotechnologies (BRAIN) Working Group*. National Institutes of Health, Washington, DC. 16 September, pp. 5–7.

interim report makes this point emphatically in its third paragraph, stating:[5]

> In analysing these goals and the current state of neuroscience the working group identified the analysis of circuits of interacting neurons as being particularly rich in opportunities, *with the potential for revolutionary advances*. [emphasis added]

The executive summary then elaborated this point:

> Truly understanding a circuit requires identifying and characterizing the component cells, defining their synaptic connections with one another, observing their dynamic patterns of activity *in vivo* during behavior, and perturbing these patterns to test their significance.

Furthermore, it stated that to achieve this level of understanding will require knowledge not only of the governance of information processing in the circuit but also between different interacting circuits in the brain as a whole.

Whilst the advisory group did not think these objectives could be easily achieved at the moment, the aim of this central thrust of the US BRAIN Initiative is not in doubt. The aim is to have a sufficient mechanistic understanding of how the brain produces behaviour to manipulate it for benign purposes. Indeed, it would be very difficult to misunderstand this central objective as the preamble to the interim report is subtitled 'The Goals of the Brain Initiative' and has only one sentence emphasised in bold type which reads:[6]

> **The challenge is to map the circuits of the brain, measure the fluctuating patterns of electrical and chemical activity flowing within those circuits, and to understand how their interplay creates our unique cognitive and behavioral capabilities.**

Yet I could find no hint in the report of the possibility that such knowledge might be subject to hostile misuse, despite the preamble later observing[7] that 'like other great leaps in the history of science... this one will change human society forever'.

This lack of awareness of the possible hostile misuse of the fruits of this intended priority research is all the more remarkable when the details of some of the research plans are set out later in the report. For example, the first section of the report has a part titled 'Foundational Concepts:

Neural Coding, Neural Circuit Dynamics and Neuromodulation', where it is noted[8] that neural coding and neural circuit dynamics are 'conceptual foundations upon which to base a mechanistic understanding of the brain', and later, I think correctly, the text goes on to suggest that just as the electrophysiological features of neurons are generalisable across many species those of neuronal circuits are also likely to have generalisable properties (see Chapter 2).

However, it is the question of neuromodulation that is of particular interest here and that deserves detailed attention. The report sets out the importance of neuromodulation very clearly:[9]

> Accompanying this rapid flow of information [in circuits] that drives cognition, perception and action are slower modulatory influences associated with arousal, emotion, motivation, physiological needs and circadian states.

and it indicates the critical role of bioregulatory chemicals in these processes, continuing as follows:

> In some cases, these slower influences are associated with specialized-neuromodulatory chemicals like serotonin and neuropeptides, often produced deep in the brain

Thus the report notes that:

> In effect, neuromodulatory modifications of synaptic efficiency can 'rewire' a circuit to produce different dynamic patterns of activity at different points in time. *The BRAIN Initiative should strive for a deeper understanding of these powerful but elusive regulators of mood and behavior.* [emphasis added]

What more, one might ask, could those with hostile intent ask for?

Looking at it from such an angle, a weaponeer might also like to be better able to target specific neuronal circuits! Here again, the results of the BRAIN Initiative might well prove very helpful. The second section of the interim report has a part titled 'Tools for Experimental Access to Defined Cell Types'. After referring to the recent remarkable development of genetic tools to target cells, the report noted:[10]

> In addition non-genetic methods could be used to deliver active agents to neurons of particular types, and would expand the range

> of possible experiments. Viruses or liposomes that contain pharmacological agents, proteins, or nano-particles might be created with antibodies that direct them to certain cell types.

The text goes on to suggest that a 'full explanation of methods for targeting genes, proteins, and chemicals to specific cell types is highly desirable' and that:

> The long term vision is the development of comprehensive general suites of tools that target expression to a brain area of interest, disseminated for broad effective use in neuroscience labs around the world.

By that time we can only hope that more thought will have been given as to how the results of such advances are to be kept out of the hands of those with malign intentions.

However, the authors of the report do not stop there. Section 6 has a sub-section with the title 'PET and Neurochemistry', which points out that a lot more could be done now to understand particular synapses and how they function in circuits. It argues that there are two challenges,[11] the first of which 'is to exploit the potential for better use of existing PET tracers that target dozens of important neurotransmitter systems and their receptor subtypes'. Importantly, from the perspective taken here in regard to future chemical and biological weapons developments, it continues:

> Within the libraries of compounds tested for therapeutic potency by the pharmaceutical industry are hundreds of compounds awaiting evaluation of their potential as *imaging* agents. Public/private partnerships under the BRAIN Initiative could unlock this potential treasure trove of compounds...as compounds for discovery of receptor function.

In the authors' opinion 'true dynamic representation of receptor occupancy and metabolism' at spatial resolution of current fMRI is a feasible mid-term goal for dopaminergic and other classical neurotransmitter systems, and the range of molecular targets and receptor subtypes should steadily grow given the cooperation of pharmaceutical companies and use of their libraries of compounds.

With knowledge of the circuits underlying behaviours of interest and of the neurochemistry of the synapses involved, the weaponeer of the future would surely also want to acquire the means of modifying the behaviour of the person targeted via manipulation of the circuits involved. Section 3 of the interim report on 'Manipulating Circuit Activity' would therefore be a welcome addition for him. The section ends by suggesting that a high priority research area for fiscal year 2014 should be to 'develop a suite of tools for circuit manipulation'. This suggestion is elaborated as follows:[12]

> To enable the immense potential of circuit manipulation a new generation of tools for optogenetics, pharmacogenetics, and biochemical and electromagnetic modulation should be developed for use in animals and eventually in human patients. Emphasis should be placed on achieving modulation of circuits and patterns that mimic natural activity.

Now, given that DARPA, the US Defense Advanced Research Projects Agency, is one of the primary institutions involved in the BRAIN Initiative,[13] it is surprising that there is no mention of the possibility of hostile misuse of such neuronal circuit modulation capabilities.

This lack applies even to the part of section 8 of the report which suggests the need to 'Consider Ethical Implications of the BRAIN Initiative'. There is definitely no mention of the Chemical Weapons Convention or the Biological and Toxin Weapons Convention. On the contrary, as if we were still living in the 1980s (see Chapter 4), it is to be a core principle of the BRAIN Initiative that it is open to all. As the text states in section 8:[14]

> In summary a core principle of the BRAIN Initiative is that new technologies and reagents should be made available across the community at the earliest possible time. This will require thoughtful development of dissemination policies by the scientific community as well as specialized support mechanisms, private/public partnerships, and training programs.

My argument here, of course, is not that such dissemination should not take place, only that there should be enough awareness of the possibility of misuse to induce a certain caution when needed.

The EU Human Brain Project

It appears that the US BRAIN Initiative originated with the idea of recording and analysing all the activity of the brain but had been honed down to a much more attainable set of objectives.[15] A February *New York Times* article on the US BRAIN Initiative noted that it was very different from the EU project that had been announced.[16] It also noted that critics felt that the EU project for constructing a supercomputer simulation of the brain (along the lines from which the US initiative was derived), was unlikely to be successful given the limited state of our current knowledge of the brain. This EU Human Brain Project (HBP) certainly appears to have a very different orientation to the US BRAIN Initiative with a much more computer/information technology central thrust. The aim is nothing less than to develop a new basis for neuroscience research. When the EU announced that it was to be one of its two flagship projects in January 2013, the press release made this obvious,[17] stating that:

> The goal of the Human Brain Project is to pull together all our existing knowledge about the human brain and to reconstruct the brain, piece by piece, in supercomputer-based models and simulations.

The project was set to begin in the closing months of 2013.

In a report to the EU the executive summary argues that this new foundation for brain research will be achieved by pursuing four goals. Two of these are not too dissimilar from those of the US project:[18]

> 1. *Data*: generate strategically selected data essential to seed brain atlases, build brain models and catalyse contributions from other groups.
>
> 2. *Theory*: identify mathematical principles underlying the relationships between different levels of brain organisation and their role in the brain's ability to acquire, represent and store information.

The other two goals, I think, reflect the much greater emphasis on information technology:

> 3. *ICT platforms*: provide an integrated system of ICT platforms offering services to neuroscientists, clinical researchers and technology developers that accelerate the pace of their research.
>
> 4. *Applications*: develop first draft models and prototype technologies demonstrating how the platforms can be used to produce results with immediate value for basic neuroscience, medicine and computing technology.

The executive summary goes on to describe briefly this integrated set of ICT platforms as: the neuroinformatics platform, the brain simulation platform, the high performance computing platform, the medical informatics platform, the neuromorphic computing platform and the neurorobotics platform.

On this basis it might seem obvious that the EU Human Brain Project with its two-and-a-half-year ramp-up phase, four-and-a-half-year operational phase and three-year sustainability phase really is very different from the US initiative. Yet both of these research proposals are drawing on the same body of knowledge and, in different ways, trying to advance the goal of creating a mechanistic (modifiable) understanding of the brain. For example, under section 8 on core principles the US BRAIN Initiative is equally enthusiastic about establishing platforms for sharing data,[19] noting that 'well-curated, public data platforms with common data standards, seamless user accessibility and central maintenance would make it possible to preserve, compare, and reanalyze valuable data sets' that have been collected.

More interestingly, from the viewpoint of this book, the following extract from the detailed discussion of the data to be collected (under goal 1 of the EU project) equally seems as if it might have been drawn from the US document:[20]

> Combined with behavioural data from elsewhere in the project, this data would provide fundamental insights into the way brain regions interact to shape perceptions and support cognition. These insights can help the project to answer new questions.

And what are these new questions? The text continues:

> Which combinations and sequences of activation of different brain regions support different forms of behaviour? How do genes and gene expression correlate with cognition and behaviour?

The mechanistic expectations can hardly have been more explicit as the list of questions goes on to ask:

> How are the building blocks of behaviour related to one another and what is their mechanistic underpinning at the molecular, cellular and circuit levels? What is the smallest network of neurons that can perform an isolated task? How does the composition of a neural microcircuit affect the computational operations it performs?

This again sounds rather similar to the US project.

The discussion of the integrative principles of cognition in the EU executive summary makes a strong a link with goals 3 and 4 of the US initiative, stating that:[21]

> Researchers should use the *Brain Simulation* and *Neurorobotics Platforms* in projects that systematically dissect the neuronal circuits responsible for specific behaviours, simulating the effects of genetic defects, lesions, and loss of cells at different levels of brain organisation and modelling the effects of drugs.

Like the US programme, the EU project lays a stress on the need for an ethical approach. Indeed, it stresses the need for researchers to be able to reflect on ethical issues, commenting under researcher awareness that:[22]

> Ethical issues cannot be reduced to simple algorithms or prescriptions: moral statements and positions always require higher-level ethical reflection and justification. From an ethical point of view, this reflection should come, not just from external 'ethical experts', but also from researchers and their leaders.

And it comments wryly, 'this kind of general reflectivity is currently not the norm and is likely to meet resistance'.

Probably because of the greater involvement of computer scientists, the EU project also notes the potential use of neuromorphic and neurorobotic technologies in drones by the military and the questions of morality and law that can arise, stating:[23]

> Neuromorphic and neurorobotic technologies have obvious applications in autonomous or semi-autonomous weapons systems, and as controllers for such systems. As has been pointed out in debates on military drones, the deployment of such systems raises delicate issues of morality and of international criminal law.

Yet it is again striking that there is no mention, in this account, of chemical or biological weapons targeted with increased precision and impact on the nervous system, or of the Biological and Toxin Weapons Convention and the Chemical Weapons Convention, from the history of which we know that there has long been a potential problem of the misuse of civilian advances for hostile purposes.

This lack of attention to misuse of new knowledge of the brain is somewhat surprising given the importance of toxicology in the practice of medical neuroscience. For example, there is the paradox of transmission involving the major neurotransmitter glutamate, in that:[24]

> activation of glutamate receptors may promote growth and survival of neurons (a trophic effect) on the one hand and cause degeneration and cell death (a toxic effect) on the other.

The toxic effect can obviously be very important in disease states such as stroke and traumatic brain injury. Furthermore, there are well-known examples of toxins that can be ingested in food that then have a neurotoxic impact. A shellfish poison called domoic acid, for example, is a very potent activator of kainate receptors in the brain and can cause death or, if the person should survive, problems with memory caused by damage to the hippocampus.

It must surely be apparent to anyone who ever listens to news broadcasts that the nature of warfare is changing in ways that could easily promote the use of chemical and biological weapons. Additionally, if neuroscientists took a wider interest in the social impacts of their work, it would not only be the use of neuroscience advances in autonomous weapons systems that would raise issues about hostile applications in the future. We turn now to these wider issues.

The changing nature of warfare

Even those of us who have little military experience recognise that the wars in which Western countries have been involved over the last twenty years are very different from the huge clashes of armoured forces that characterised earlier wars in the twentieth century. However, the British general, Rupert Smith, took a much more dramatic view of the change in his book, *The Utility of Force: The Art of War in the Modern World*. In his view such war no longer exists.[25]

Smith sets out six main characteristics of warfare in current post-Cold War times, such as, the sides are mostly not states and new uses being found for old weapons. However, most importantly here:

> We fight among the people, a fact amplified literally and figuratively by the central role of the media: we fight in every living room in the world as well as on the streets and fields of a conflict zone.

As *this* book is being written, it is clear that the western public has lost its enthusiasm for engaging directly by sending their armed forces on the ground into such messy conflicts.

Yet such conflicts will continue and, as Perry Robinson has cogently argued, participants could find new uses for chemical and biological weapons.[26] As we saw in Chapter 3, most recent concern has focused on the danger that developed states, in seeking new incapacitating chemical agents to use in such conflicts – for example, the use of fentanyl derivatives to break the 2002 Moscow theatre siege – may place the prohibition of chemical and biological weapons in jeopardy. However, if we consider the possible evolution of the biotechnology revolution as set out by Meselson (see the end of Chapter 3), we cannot believe that is all that might be of future concern. As Perry Robinson states:[27]

> if a new molecule is discovered that can exert novel disabling effects on the human body at low dose, attempts to weaponize it may well ensue.

Or:

> if a hitherto unknown molecular pathway serving a process of life comes to be identified, chemical agents capable of interfering with that pathway might also become identifiable and then form the basis for a novel weapon.

It is important to follow this line of thought to its logical endpoint.

In the classic Stockholm International Peace Research Institute (SIPRI) 1970s study, *CB Weapons Today*,[28] the section on incapacitating agents notes that in the early stages of the US incapacitants programme there seemed to be many possible mechanisms of incapacitation that new agents might be developed to exploit. These included hypotension, emesis, disturbance of body temperature and balance, temporary blindness and uncontrollable tremors. However, we now know that the scientists involved during the 1950s and 1960s were struggling with an extremely limited understanding of the neuronal circuits and receptor systems that they were trying to disrupt.

What we need to think about now is what current and future weaponeers would want to achieve and how the revolution in biotechnology might facilitate their objectives. So we might well ask:

What range of functions is it now possible to envisage disabling?
What precise targets involved in the operation of such functions can be targeted?
Can new agents be developed to attack such targets in specific ways without affecting other targets in other functional systems?
Can such chemical and biological agents be easily produced and stored in adequate quantities for effective use?
Can such agents be dispersed over small, medium or even very large areas in such a way that they will enter the brains of people in that area and effectively cause the disabling effect desired?

From what we know of earlier work it might, for example, be expected that a weaponeer would consider disabling human special senses (causing temporary blindness), interfering with human homeostatic systems (blood pressure, balance or consciousness) and some aspects of higher order functions (such as cognition, as was the result of using anticholinergic glycollates like BZ). Given our increasing understanding of mental illnesses, is it not likely that causing depression, anxiety and so on might also be envisaged? And with the recent growth in understanding of social neuroscience, disruption of much normal interactive behaviour might also seem to be a viable objective (Table 5.3). Add to all that the possibility that the weaponeer might have in mind finding new means of interrogation and torture,[29] not just military or crowd control objectives, and it is not difficult to see how the dangers multiply if society is unable to prevent the future hostile misuse of the biotechnology revolution.

Table 5.3 Functions that might be disabled

Special senses	e.g. vision
Homeostatic systems	e.g. blood pressure, balance, consciousness
Higher order capabilities	e.g. cognition, memory
Mental health	e.g. depression, anxiety
Social interaction	e.g. trust

It is to growing concerns about such distortions of the revolution in civil neuroscience that we turn in the next chapter.

References

1. Editorial (2013) Drop medicine not bombs: Military force is not the way to help Syrians threatened by gas attacks. *New Scientist*, 31 August, p. 5.
2. Markoff, J. (2013) Obama seeking to boost study of human brain. *The New York Times*, 17 February.

3. Advisory Committee to the NIH Director (2013) *Interim Report: Brain Research through Advancing Innovative Neurotechnologies (BRAIN) Working Group*. National Institutes of Health, Washington, D.C. 16 September.
4. Dando, M. R. (2007) Scientific outlook for the development of incapacitants, pp. 123–147 in A. M. Pearson, M. I. Chevrier and M. Wheelis (Eds) *Incapacitating Biochemical Weapons: Promise or Peril?* Lanham, MD: Lexington Books.
5. Reference 3, p. 4.
6. ibid, p. 8.
7. ibid, p. 9.
8. ibid, p. 11.
9. ibid, p. 12.
10. ibid, p. 19.
11. ibid, p. 42.
12. ibid, p. 32.
13. See reference 2.
14. Reference 3, p. 52.
15. Shen, H. (2013) *Neurotechnology: BRAINstorm*. ***Nature*** *News Feature*. Available at <http://www.nature.com/news/neurotechnology-brain-storm-1.141057>. 31 December 2013.
16. See reference 2.
17. HBP (2013) *The Human Brain Project Wins Top European Science Funding*. Press release, Human Brain Project, embargo Monday, 28 January, 11.15 (Central European Time).
18. The Human Brain Project (2012) *A Report to the European Commission*, April, p. 9.
19. Reference 3, p. 50.
20. Reference 18, p. 30.
21. ibid, p. 12.
22. ibid, p. 54.
23. ibid, p. 92.
24. Brady, Scott T. and Seigal, G. J. (2012) *Basic Neurochemistry: Principles of Molecular, Cellular and Medical Neurobiology*. Oxford: Academic Press. (p. 363).
25. Smith, R. (2006) *The Utility of Force: The Art of War in the Modern World*. Penguin Books, London. (p. 1).
26. Perry Robinson, J. P. (2008) Difficulties facing the Chemical Weapons Convention. *International Affairs*, **84** (2), 223–239.
27. ibid, p. 227.
28. SIPRI (1973) *The Problem of Chemical and Biological Warfare: Volume II CB Weapons Today*. Almquist and Wiksell, Stockholm. (pp. 298–300).
29. Bowden, M. (2003) The dark art of interrogation. *Atlantic Monthly*, October, Available at <http://www.theatlantic.com/doc/200310/bowen>. 17 April 2004.

6
Novel Neuroweapons

Introduction

In 2007 a set of lecture powerpoint slides and speaking notes appeared on the internet.[1] They were titled 'Protecting our National Neuroscience Infrastructure: Implications for Homeland Security, National Security and the Future of Strategic Weapons'. What was interesting for a start was that the lecture was written by Dr. Robert E. McCreight who has spent 35 years in the United States State Department working on global security, arms control, biowarfare, treaty verification and other related issues.[2]

The lecture raised some of the questions that have been extensively discussed in this book. For example:

> Have we adequately analyzed and discussed the dual-use implications of neuroscience, particularly its various military applications, and the extent to which operational safeguards and societal controls are needed to manage or control its most destructive weapons outcomes or debilitating systems?

and:

> What assurances do we have that all future neuroscience research both here and globally, not just 80% or 90%, is devoted exclusively to helping brain functions, extend mental health, find cures for neurological diseases and enable better mechanisms for understanding the principles and operations of neurobiology?

But it also raises the issue of paradigm-changing technological development, noting that 'effective neuroscience weapons could significantly affect and redefine existing concepts of strategic warfare'. In this regard McCreight also pointed out that 'policymakers, legislators, scientists and the public must insist on more information and improved transparency about activities'.

Perhaps coincidentally, the US Quadrennial Defense Review the previous year had suggested that one such 'disruptive threat' – something unexpected that could radically change the current balance of military power – could be developments in biotechnology.[3] We will return to that point shortly as it is a recurrent theme in discussions on the potential impact of biological warfare. Yet it is the point about the necessity for transparency and public debate on what needs to be done that should be particularly noted because in 2014, seven years later, transparency and debate are *still* absent and very likely to remain so for some years to come.

That lack of debate was criticised in an editorial early on in the new century by *The Economist*,[4] which argued that there was a major discussion of the implications of advances in molecular biology and genetics, but 'if you want to predict and control a person's behaviour, the brain is the place to start'. The editorial pointed out that no such discussion about the implications of advances in neuroscience was taking place. An accompanying article on the ethics of brain science argued that concerns about genetics fell into three categories:[5]

> first, how much screening should be allowed for certain genetic traits; second, who should have access to such information; and third, what will happen when those traits can be modified at will.

The article went on to suggest that concerns about the implications of neuroscience fell into precisely the same three categories. And, of course, the last category includes the possibility of hostile misuse.

Before looking at how discussions on the possible misuse of neuroscience have developed more recently, particularly with respect to chemical and biological weapons, it is useful to consider the broader issue of the possible strategic implications of the development of such novel chemical and biological weapons designed specifically, on the basis of new neuroscience research, to attack the brain.

Chemical and biological weapons as a disruptive threat

There are numerous official and semi-official reports in the open literature about the future of chemical and biological weapons. One recurrent theme is that a crucial difficulty for a non-state actor wishing to use such weapons is *delivery*, particularly of biological weapons. A technical report from the mid-1990s pointed out that:[6]

> Weaponization, mating the agent with a disseminating device and a delivery system poses special problems that may be difficult to overcome... Ballistic missile delivery will require a sufficiently demanding aerosol generator that only the most sophisticated program is likely to be able to develop.

Such official opinions are likely to be based on other studies within the closed literature.[7]

Numerous studies by academic and other analysts are also available in the open literature. A recurrent theme in these studies is the recognition of the importance of military (or police) forces' interest in such weapons. In a large multi-author study of biological weapons since 1945 it was concluded that:[8]

> The major states that ended on the winning side after World War II had developed BW programs because such weapons were seen to be potentially important militarily for retaliation in kind, and they continued or restarted them for the same reasons...The two states definitely known to have begun offensive programs later in the century also had military reasons for their programs.

Other authors have argued that in present circumstances weaker states might also view such weapons as a useful deterrent against attack by more powerful states.[9,10]

Another obvious theme in the open literature is the sheer diversity of the threat. As Jonathan Tucker concluded in 2010:[11]

> The CW threat is multifaceted, encompassing military-grade agents, novel incapacitating agents, and toxic industrial chemicals. Moreover, in a world of globalized, flexible chemical manufacturing, countries may decide to hedge their bets by acquiring a standby capability to produce CW agents in a crisis or war.

Tucker went on to note that such a latent capability would be very hard to detect before it became operational at short notice.

Such diversity of the chemical and biological weapons possibilities is widely acknowledged[12] and within that diversity so is the possibility of the development of new forms of non-lethal weapons:[13]

> the persistence of insurgent and terrorist threats to many states could increase those states' interest in non-lethal and low-lethal chemical agents. As the United States once again learned in Iraq and Afghanistan, collateral damage to the local population by counter-insurgency and counterterrorism operations can foster and sustain local support for the insurgents and terrorists.

Of course, there is no reason to believe that such weaponry would remain non-lethal[14] if states decide to go down that road, as is obvious from what we know of the Soviet bioweapons programme.[15] Against that background, what are we to make of recent concerns about novel bioweapons targeted at the human nervous system?

Neuroscience and novel neuroweapons

In early 2014 the United States National Academies published a large study entitled *Emerging and Readily Available Technologies and National Security: A Framework for Addressing Ethical, Legal, and Societal Issues*. The preface began by explaining that the study was carried out following a request from the Defense Advanced Research Projects Agency (DARPA) in 2010 and a footnote explains the title in this way:[16]

> DARPA's original charge to the committee used the term 'democratized technologies' rather than 'emerging and readily available technologies'. Democratized or, equivalently, emerging and readily available technologies are those with rapid rates of progress and low barriers to entry.

The committee decided not to use the term 'democratized technologies' because they thought that it might be misunderstood. Others might feel that 'emerging and readily available technologies' is a much clumsier expression, but what is clear is that the committee was investigating what Meselson[17] had described as technologies that have 'the potential to place mass destructive capabilities in a multitude of hands'.

In that regard it should be noted that the committee's preface to its report goes on to state:[18]

> the committee received input on specific emerging and readily available technologies including *information technology, neuroscience, prosthetics and human enhancement*, synthetic biology, cyber weapons, robotics and autonomous weapons, and *nonlethal weapons*. [emphases added]

Neuroscience is, in fact, one of three foundational technologies dealt with in chapter 2 of the report, along with information technology and synthetic biology.

Chapter 2.3 begins by rehearsing various definitions of neuroscience before stating that modern neuroscience:[19]

> is thus an interdisciplinary field that combines new knowledge of molecules, cells, neural circuits, and cognition; is allied with clinical medicines; and uses methodologies of mathematics, molecular biology, genomics, neuroendocrinology, neuroimaging, and social and behavioural sciences.

It goes on to argue that the annual growth in the number of publications on neuroscience indicates the maturity of the field. Publications are shown to have increased by a factor of between eight and ten times over the past 20 years. A similar rapid growth can be seen in the membership of the Society for Neuroscience from 18,976 in 1991 to 42,576 in 2011.

The committee sees two classes of possible military applications: those that help to restore normal functions and those that change normal functions. In the first class are advances in neuroscience which help soldiers who have suffered traumatic brain injuries. In the second class there are seen to be two sub-classes: those that enhance normal functions, such as neuroscience-based applications like brain-machine interfaces that assist in the operation of equipment; and those that diminish normal functions. It is the second category which is of most interest here and, as the report puts it:[20]

> Much more controversial from an ELSI [ethical, legal, social implications] standpoint are other proposals suggesting that false human memories can be created and different emotional states induced (e.g. reduced or increased fear, feelings of anger or calm) and that degrading the performance of adversaries in military contexts may be possible.

The report goes on to discuss recent work on cognitive enhancement and brain-computer interfaces, and deception, detection and interrogation, before turning to performance degradation.

In regard to degradation, the committee is quite clear about what could be of concern:[21]

> At present, the primary focus for such efforts to support military missions and law enforcement goals – as well as applications in

> areas such as counterterrorism or counterinsurgency where the lines between the two domains are often blurred – is on so-called incapacitating chemical agents (ICAs).

The report concludes that a number of recent technical reviews have shown that at present there is no ICA that can be used safely in the field. However, it also notes the ongoing relevant advances, for example in neuropharmacology, bioregulatory chemicals, delivery of drugs to the brain (particularly via aerosols), and the use of nanotechnology for targeted delivery of agents to the brain.

Chapter 3 of the report discusses military applications. In the first section (3.1) the issue of robotics and autonomous systems is discussed. For the purposes of the report[22] an autonomous system 'refers to a standalone computer-based system that interacts directly with the physical world'. So such systems have sensors to input information and actuators to produce output. Obviously, such systems can fail because of errors in programming or in encountering situations for which their design is inadequate. There are already military autonomous systems that are equipped with lethal capabilities. The report lists, for example, the SWORDS platform used in Iraq and Afghanistan which can carry machine guns.

As the report points out:[23]

> Neuroscience may be an enabling technology for certain kinds of autonomous systems. Some neuroscience analysts believe that neuroscience will change the approach to computer modeling of decision making by disclosing the cognitive processes produced by millions of years of evolution, processes that artificial intelligence has to date been unable to capture fully. Such processes may become the basis for applications such as automatic target recognition.

As we saw in the previous chapter, the huge increases in funding for brain research is intended not only to produce a better mechanistic understanding of the central nervous system but simultaneously to radically alter information technology systems.

The advantages that deployment of such superior information technology may give to military forces is well understood[24] as, to some extent, are the limits to advances in neuroscience.[25] Whilst this is not the subject of the current text, it is impossible to avoid the grave threats to current international law posed by lethal autonomous systems[26] and the need for new developments in international law to deal with the problem.[27]

Sections 3.2 on prosthetics and human enhancement and 3.3 on cyber weapons in the report clearly also raise difficult questions, but it is section 3.4 on, so-called, non-lethal weapons which is of most concern here. The authors use the US Department of Defense definition of such weapons:[28]

> weapons designed and primarily employed to incapacitate targeted personnel or materiel immediately, while minimizing fatalities, permanent injury to personnel, and undesired damage to property in the targeted areas or environment.

They also note that these weapons go under a variety of other designations, such as 'less lethal, less than lethal, prelethal and potentially lethal', which perhaps indicates some of the difficulties in the safe use of such weapons on the field.

The text goes on to show how, as has been pointed out many times in the past,[29] a very wide range of weapons systems – kinetic, optical, acoustic, directed-energy, electrical and cyber – could come under the designation of non-lethal weapons. However, we are interested here predominantly with biological and chemical agents which:[30]

> may target neurological functions to incapacitate people, repel them (e.g. with a very obnoxious odor), or alter their emotional state (e.g. to calm an angry mob, to induce temporary depression in people).

The authors also make two general points about non-lethal weapons. The first is that proponents of such weapons see them as having uses in many of the situations that military, police and other forces are likely to face in coming decades. These include peacekeeping, humanitarian interventions, civil and military law enforcement and control of violent criminals. Nevertheless, it is clear that non-lethal weapons can also be used to supplement lethal force and the report quotes NATO doctrine on this point:

> Non-lethal weapons may be used in conjunction with lethal weapon systems to enhance the latter's effectiveness and efficiency across the full spectrum of military operations.

It has long been a concern that a non-lethal system may be used in order that lethal force can then be deployed more effectively, say, to flush people out of cover.

The second point is that the report states clearly that NATO doctrine does not oblige its forces to use non-lethal means before lethal means. The relevant doctrine is reproduced as: 'Neither the existence, the presence, nor the potential effect of non-lethal weapons shall constitute an obligation to use non-lethal weapons, or impose a higher standard for, or additional restrictions on, the use of lethal force.'

Chemical incapacitants

This report on emerging and readily available technologies did not come out of the blue. In regard to chemical incapacitants, a series of similar open reports has been produced in the United States since the beginning of the new millennium (see Table 6.1). Since the Second World War, and the discovery of chemical drugs that could help some people affected by mental illnesses, interest in incapacitating chemical agents affecting the central nervous system has waxed and waned amongst police and military forces. Sometimes this interest has been reflected in the open literature (for earlier reviews see [31,32,33,34]), but it is clear that until the genomics revolution began to reveal the full complexity of the neurotransmitter and neuroreceptor systems in the brain progress was necessarily limited in both the search for better drugs and for incapacitants.

The first report noted in Table 6.1, *The Advantages and Limitations of Calmatives for Use as a Non-Lethal Technique*,[35] was produced by the College of Medicine of Pennsylvania State University at a time when there appeared to be interest in non-lethal chemical agents in the United States. Calmatives are defined in the report[36] as 'pharmacological compounds (or agents) producing a calm or tranquil state upon administration', and the report aimed to provide a comprehensive survey of the medical literature in order to identify agents that might serve that purpose. What is of interest here is not simply the range of agents identified but more the identification of the receptor sub-types that were seen to be the target of these agents. As the report emphasised:[37]

Table 6.1 Some studies in the United States since 2000 involving incapacitating chemical agents

1. *The Advantages and Limitations of Calmatives for Use as a Non-Lethal Technique* (2000).
2. *Emerging Cognitive Neuroscience and Related Technologies* (2008).
3. *Opportunities in Neuroscience for Future Army Applications* (2009).
4. *Human Performance Modification: Review of Worldwide Research with a View to the Future* (2012).

> There is a need for non-lethal techniques with a high degree of specificity, selectivity, safety and reversibility to avoid producing a lasting impairment to the subject(s)

The report's authors concluded that they had identified several drug classes and individual drugs suitable for immediate consideration.

Like many others, they had not given proper consideration to the difference between using a drug in a medical setting – where the dose can be controlled and the subject carefully monitored by qualified staff – and the operational use of the same agent in the field. However, what is interesting is the range of agents and receptors considered, for example dopamine receptor agonists and opioid receptor agonists, and the conclusion that two of the drug classes – benzodiazepines and alpha 2 adrenoreceptors agonists – were prime candidates for immediate consideration. It is ironic, in view of the events in Moscow where a terrorist hostage situation was dealt with by the use of fentanyl derivatives a couple of years later, that the cover of the report also bears a diagram of the chemical structure of the opioid fentanyl.

The second report, listed in Table 6.1 was published in 2008 by the United States National Academies Press. As the summary of *Emerging Cognitive Neuroscience and Related Technologies* shows, the committee was tasked by the Defense Intelligence Agency [38] 'to identify areas of cognitive neuroscience and related technologies that will develop over the next two decades and that could have military applications'. Thus can there be no doubt that the content of this report and its conclusions are of serious military interest.

The word 'cognitive' in the report is used to refer to the processes of human information processing, whilst 'neuroscience' has a similar broad meaning to include the study of 'the central nervous system (e.g. brain) and somatic, autonomic, and neuroendocrine processes'.[39] The weighty report has five chapters and a number of appendices, but Chapter 5 on 'Potential Intelligence and Military Applications of Cognitive Neuroscience and Related Technologies' is of most interest here. In this chapter the report suggests there are three 'markets' that need to be considered in estimating future trends:[40]

> *Health*. Customers are seeking help in addressing mental illness, brain disease and injury;
>
> *Enhancement*. Customers do not possess a diagnosable neurological disorder but are seeking some cognitive performance advantage or want to prevent a probable decline; and

> *Degradation*. Customers seek advantage by degrading, temporarily or permanently, the cognitive abilities of others.

The authors obviously see the first market as large, unfulfilled and therefore likely to grow significantly. The second they see as growing but largely underground, and the third as completely underground.

However, in the authors' *Finding 5.2* it is noted that:[41]

> Neurotechnology products may be dual-use. Products intended for the health market can be used in the enhancement and degradation markets. Additionally, the product life cycle can be shunted because of the nature of the enhancement and degradation markets.

On this basis they make a number of speculations about the degradation market (Table 6.2), and then suggest a number of possible threats that might arise for US warfighters.[42]

The description of the degradation market given in the report can clearly be seen to have positive feedback loops. As shown in Table 6.2, fears that cognitive weapons might be developed (Stage 1) lead to investigation of means to deal with possible weapons (Stage 2). This in turn leads to the development of novel weapons to test against and a rapid escalation of measures and countermeasures (Stage 3) and then to function creep as agents become used in different fields, such as

Table 6.2 The degradation market

Stage 1
'The fear that this [degradation neurotechnology] approach to fighting war might be developed will be justification for developing countermeasures to possible cognitive weapons.'
Stage 2
'Tests would need to be developed to determine if a soldier had been harmed by a cognitive weapon. And there would be a need for a prophylactic of some sort.'
Stage 3
'If a particularly effective degradation product is developed that has few side effects, escalation of this market will be self-fulfilling.'
Stage 4
'The concept of torture could also be altered by products in this market.'

Source: Modified from Committee on Military Intelligence Methodology For Emergent Neurophysiological and Cognitive/Neural Science Research in the Next Two Decades (2008) *Emerging Cognitive Neuroscience and Related Technologies*. Washington D.C: National Academies Press, (p. 133).

torture (Stage 4). If this seems far-fetched it is important to note that the chairman of the committee that produced this report is on record as stating:[43]

> The study does not address arms control issues directly, yet implicit is ample evidence that in the next 20 years, the pace of development of neuroscience technologies related to the military and intelligence communities will swamp traditional arms control measures.

He later moderated his view because he felt that nongovernmental bodies might play a role in influencing rates of research,[44] but others might find his reasoning on this point less than convincing because he is a practising neuroscientist rather than a sociologist.

The report goes on to outline some possible threats in the main text. For example:[45]

> One type of identifiable threat might be the development of antagonists for drug entities that currently have no antagonists... Such an antagonist could allow adversaries to protect their own warfighters against an agent that is widely dispersed (for example, in gas, in aerosol, in drinking water, or by high altitude delivery).

The report goes on to emphasise the diversity of the threats:

> An adversary could train warfighters to operate under the influence of chemical agents that ordinarily disrupt performance or could modify warfighters to resist such agents. Such resistance could be conferred by means of changes in genetics, physiology, training, pharmacology or psychology.

Moreover, none of this is to be considered as possible only in the distant future. As the text points out 'new technologies, particularly nanotechnologies, will enable unparalleled access to the brain'. Also, 'increased experimentation with neuropeptides will have profound implications for the neuropharmacological modulation of behavior'. A series of charts then illustrate some of these possibilities (Table 6.3), and also suggest some of the indicators that might signal that such developments are taking place.

The third report listed in Table 6.1, *Opportunities in Neuroscience for Future Army Applications*,[46] was produced in 2009 by a committee chaired by Floyd Bloom, a very well-known neuroscientist. In his preface

Table 6.3 Some potential development areas of concern

Chart 5.1	*Use of Neuropharmacological Agents as Incapacitants* 'Aerosols of opioids serve as excellent incapacitants.'
Chart 5.2	*Nanotechnologies or Gas-Phase Technologies that Allow Dispersal of Highly Potent Chemicals over Wide Areas* 'technologies that could be available in the next 20 years would allow dispersal of agents in delivery vehicles that would be analogous to a pharmacological cluster bomb or a land mine.'
Chart 5.3	*Technologies for Highly Potent Blood Pressure Agents or Sensory Specific Pharmacological Targeting* 'Existing pharmacological agents could be used in a nefarious way. An example would be currently used alpha blockers, that would work quickly to drop blood pressure if delivered in high doses.'
Chart 5.4	*Drug-Delivery Systems Applied to the Blood-Brain Barrier* 'New nanotechnologies have allowed molecular conjugation or encapsulation that may permit unprecedented access to the brain.'

Source: Modified from Committee on Military Intelligence Methodology For Emergent Neurophysiological and Cognitive/Neural Science Research in the Next Two Decades (2008) *Emerging Cognitive Neuroscience and Related Technologies*. Washington D.C: National Academies Press, (pp.136–139).

to the report Bloom stressed the rate of change in neuroscience and neurotechnology:[47]

> I believe, as many others do, that despite the almost constant growth of neuroscience over the past four decades, the future of neuroscience applications will grow at a rate that has not been seen since the birth of microprocessor-based personal computers.

Bloom continued by noting that the report was being published against this background of rapid and probably accelerating change.

The report ranges very widely, with chapters on training and learning, optimising decision-making, sustaining soldier performance and improving cognitive and behavioural performance. It also has a list of high-payoff research opportunities including, under the sub-title 'Pharmaceutical Countermeasures to Performance Degradation,' the following Recommendation 7:[48]

> The Army should establish relationships with the pharmaceutical industry, the National Institutes of Health, and academic laboratories to keep abreast of advances in neuropharmacology, cellular and molecular neurobiology, and neural development and to identify

> new drugs that have the potential to sustain or enhance performance in military-unique circumstances.

This looks as if such a search for countermeasures would fit exactly into Stage 1 and Stage 2 of the dangerous process illustrated in the description of the Degradation market in Table 6.2.

A similar view might well be taken of Recommendation 9 which states:

> The Army should support research on novel mechanisms for noninvasive targeted delivery of pharmacological agents to the brain and nervous system in the course of medical interventions to mitigate the adverse effects of physical injury to the brain or another portion of the nervous system.

Whilst the results of such research for people with injuries could be of benefit, the dual-use potential should surely also be borne in mind and guarded against.

If anyone needed reminding that benignly intended research in neuroscience can be turned to other uses, it was made abundantly clear in the fourth report listed in Table 6.1. In late 2011 the US Army requested the National Research Council to set up a committee to review modern research around the world on human performance modification. The committee's report, *Human Performance Modification: Review of Worldwide Research with a View to the Future,*[49] was published in 2012. The report summary begins by making it clear that advances in this area are of increasing military concern[50] and can include 'the application of nanotechnology as a drug-delivery mechanism or in an invasive brain implant'. Additionally, whilst it sees the literature in this field as usually addressing the question of enhancement, 'another possible focus is methods that degrade performance or negatively affect a military force's ability to fight'.

Most of the report ranges widely over issues that are not of immediate concern here, but one example of technologies for degrading human performance illustrates just how simple incapacitation can be. The committee reports that it identified a *class* of technology designed to degrade performance and gave as an example[51] Japanese researchers who have developed a device, using commercial off-the-shelf components, to interfere with and prevent speech production.

The text goes on to argue that use of such a device could lead to serious consequences by preventing spoken commands, instructions, or assurances intended for friendly or enemy troops, or civilians.

The report refers to the Japanese researchers' paper titled *Speech jammer: a system utilizing artificial speech disturbance with delayed auditory feedback*.[52] As these researchers explain:

> In general, human speech is jammed by giving back to the speaker their own utterances at a delay of a few hundred milliseconds. The effect can disturb people without any physical discomfort, and disappears immediately [when they] stop speaking. Furthermore, this effect does not involve anyone but the speaker.

The paper continues by describing two prototype systems that combine a direction-sensitive microphone and a direction-sensitive speaker which together allow the speech of a particular person to be disturbed. This example clearly demonstrates how fundamental knowledge of the operational characteristics of the human brain can be used for hostile purposes. But, at a more general level, this fourth report listed in Table 6.1, with its emphasis on worldwide research, reminds us that work that could be misused, and concerns about such potential misuse, are not confined to the United States. It is to one further example that we now turn, from the Royal Society of the United Kingdom, to examine in more detail some possible chemical agents that might be misused.

Brain waves module 3

Publications raising concerns about misuse, at least in part related to advances in neuroscience, have appeared in the open literature sporadically since the turn of the millennium[53,54,55,56,57] but what is interesting about the 2012 UK Royal Society Report, *Neuroscience, conflict and security*,[58] is its exclusive concentration on advances in neuroscience and particularly their potential misuse in novel incapacitating agents.

Along with other technical analyses at that time,[59] the Royal Society report did not believe there was an incapacitating agent that could be used safely in the field, stating:

> as this review illustrates, the feasibility of developing an incapacitating chemical agent and delivery system combination that is safe (i.e. has a low risk of lethality) is questionable.

Nevertheless, there is a wide range of pharmaceutical agents that some might consider as potentially of use and the report examines opioids, benzodiazepines, alpha 2 adrenergic agonists, neuroleptic anaesthetics and bioregulators in general. During the Cold War many different means of chemical incapacitation were examined. For example, in the late 1960s the United States tested an F4/AB45Y-4 incapacitant weapon system in the Pacific. A jet plane disseminated an aerosol of Agent PG in a line and the spread of the agent was monitored downwind. The net result was this:[60]

> A single weapon was calculated to have covered 2400 square km, producing 30 percent casualties for a susceptible population under test conditions.

That area is equivalent to about 926 square miles, roughly twice the size of metropolitan Los Angeles. Agent PG is Staphylococcal Enterotoxin type B (SEB)[61] which is a superantigen[62] that is likely to make people ill within a few hours and to make them feel ill in varying degrees for several days.[63] However, as the UK Royal Society report noted:[64]

> Many different forms of incapacitation were investigated during the Cold War...but with increasing emphasis on rapid action and short duration of effect, contemporary interest has tended to focus on sedative-hypnotic agents that reduce alertness, and, as the dose increases, produce sedation, sleep, anaesthesia and death.

Clearly in order to deal with something like a hostage situation, such a rapid action/short duration effect would be preferable – if it was possible to achieve in the field. Here we briefly review ongoing work on one of the four groups of chemicals of this type identified in the Royal Society report.

Opioids

Opioids can be divided into four groups, natural opiates like morphine, which are derived from the opium poppy and which have long been used for pain relief; semi-synthetic opioids like heroin, which are derived from natural opiates; fully synthetic opioids, such as fentanyl and its derivatives that are made by chemical synthesis; and endogenous opioid peptides (strings of amino acids) that are produced naturally in the body.

Opioids are standardly used in medicine for pain relief by medically trained people. However, care has to be taken by medical practitioners because, like all drugs, opioids can have side effects As is well known:[65]

> it is perhaps respiratory depression that remains the main hazard of opioid use, uppermost in the minds of nurses and physicians because of the obvious risk of fatal outcome.

It has been understood since the 1960s that the endogenous opioids act through a number of opioid receptors and natural, semi-synthetic and fully synthetic opioids do likewise. Indeed, it is well-known that the respiratory depression is caused largely by the interaction of these agents with μ-opioid receptors. Medical chemists have also managed to modify the structure of fentanyl to produce opioids such as remifentanil and carfentanil, which are much more potent than fentanyl. Carfentanil is one hundred times more potent than fentanyl and ten thousand times more potent than morphine. Standardly, it is only used by veterinarians to incapacitate large wild animals. Fortunately, the action of the opioid can be reversed by use of naloxone if it is given in good time, as was demonstrated when a vet was accidentally contaminated with carfentanil.[66]

According to a report soon after the breaking of the 2002 Moscow theatre siege, the Russian Health Minister stated that 'a fentanyl derivative was used to neutralize the terrorists' and that the gas 'cannot by itself be called lethal'.[67] But 127 (16 per cent) of the 800 hostages in the theatre died and more than 650 of the survivors required hospitalisation, according to a 2003 medical analysis. However, the authors remained uncertain as to the exact agents used by the Russian Special Forces. The nature of the agents used became clearer when a paper by the UK's Defence Science and Technology Laboratory at Porton Down was published in 2012.[68] These scientists reported on analyses of clothing and medical samples taken from two British survivors (Casualties 1 and 2) and medical samples taken from a third survivor. The two British survivors reported that:

> From first spotting the aerosol to being overcome by it took '10 to 30 seconds' for Casualty 1 and 'at least 30 seconds' for Casualty 2. Neither casualty saw the assault team enter the theatre.

Both British survivors were near the exit door and were amongst the first casualties removed. Casualty 1 woke up in hospital and Casualty 2 drifted in and out of consciousness on the way to hospital. The Russian doctors did not know the nature of the incapacitating agents, which did not help their rescue efforts. Nevertheless, samples arrived at Porton

Down on 28 October, two days after the ending of the siege and both British survivors were well enough to be interviewed by UK police in early November.

Through the use of very sophisticated and sensitive analytical techniques the scientists at Porton Down were able to conclude that:

> This case report provides evidence from the analysis of clothing from two British survivors and urine from a third survivor with a Russian name, that the aerosol comprised a mixture of two fentanyls, carfentanil and remifentanil.

They also speculated that carfentanil may have been mixed with larger quantities of the less potent, shorter-acting remifentanil in order to reduce the number of casualties that would have resulted from the use of carfentanil alone. Nevertheless, they made it clear that:

> Both carfentanil and remifentanil have narrow safety margins, meaning that potentially fatal side effects, including respiratory depression, can occur at doses only slightly higher that those that impart medical benefits.

In an operation in the field like the Moscow siege there was no means of controlling the concentration of the agents taken in by any victim or of knowing the specific response of any particular person – young, old or ill, or adult and healthy. The result of the fentanyl derivatives affecting differing sets of opioid receptors, including those producing respiratory depression and besides those producing sedation, was an inevitable likelihood of causing casualties and fatalities.

It may seem that the possible use of fentanyl derivatives ends there. However, that would be to ignore ongoing advances that could become relevant. The same type of receptor can operate in different functional circuits, thus:[69]

> opioid-induced respiratory depression and analgesia are inextricably linked by their mediation through the μ-opioid receptor. Reversal of opioid-induced respiratory depression by naloxone, therefore, may lead inevitably to the loss of analgesia, which creates difficulties in patient care.

In such circumstances it is logical to try to find other ways of dealing with respiratory depression. One possible solution is to enhance the output of the centres in the brain stem that drive respiration through the use of other drugs, that is, to overcome the depression caused by the opioid without the use of naloxone. This approach seems to be a possibility:[70]

> New treatments and/or approaches to prevent opioid respiratory depression without affecting analgesia have led to the experimental application of new agents such as serotonin agonists, ampakines, and antibiotic minocycline. There are other promising agents available that deserve study…that have a stimulatory effect on breathing due to an action within central respiratory pathways.

Should such approaches work well in medical practice, what might be the implications for dual-use in the development of enhanced opioid incapacitants?

Conclusion

Clearly, whilst it is not well-known outside specialist communities, there has been increasing concern about the hostile application of advances in neuroscience. In the next chapter we take a longer-term view of where such developments might lead in coming decades.

References

1. McCreight, R. (2007). *Protecting Our National Neuroscience Infrastructure: Implications for Homeland Security, National Security and the Future of Strategic Weapons*, Institute of Crisis, Disaster and Risk Management, January 2007. Available at <http://www.authorstream.com/Presentation/Nastasia-54479-McCreight-Protecting-National_Neuroscience_Infrastructure>.
2. McCreight R. E. Available at <http://www.zoominfo.com/p/Robert-McCreight/1173822452>. 07 February 2014.
3. Dupont D. G. (2006) Disruptive threats: The Pentagon tries to know the unknown. *Scientific American*, June, 9–10.
4. Editorial (2002) The future of mind control. *The Economist*, 25 May, p. 11.
5. Science and Technology (2002) The ethics of brain sciences: open your mind. *The Economist*, 25 May, 93–95.
6. U.S. Army Medical Research Institute of Infectious Diseases (1994) *Biological Weapons Proliferation: Technical Report*. Contract No. DNA-MIPR-90–715, Defense Nuclear Agency, Alexandria, VA, USA. (p. 6).
7. Grim, B. S. (1981) *Biological Agent Delivery by ICBM*. Project No. 1L162706A553-TA3–7, U.S. Army Dugway Proving Ground, Dugway, UT, USA.
8. Wheelis, M., Rózsa, L. and Dando, M. R. (Eds) (2006) *Deadly Cultures: Biological Weapons Since 1945*. Harvard: Harvard University Press, (p. 360).

9. Martin, S. B. (2002) The role of biological weapons in international politics: The real military revolution. *Journal of Strategic Studies*, **25** (1), 63–98.
10. Preston, T. (2007) *From Lambs to Lions: Future Security Relationships in a World of Biological and Nuclear Weapons*. Lanham: Rawman and Littlefield.
11. Tucker, J. B. (2010) The future of chemical weapons. *The New Atlantis*, Fall 2009/Winter 2010, 3–29.
12. Davie, J. A. and Schneider, B. R. (2002) *The Gathering Biological Warfare Storm*. USAF Counterproliferation Center, Maxwell Air Force Base, Alabama, USA.
13. Caves, J. P. (2010) Future foreign perceptions of chemical weapons utility. *WMD Proceedings*, October, 1–4. Center for the Study of Weapons of Mass Destruction. Available at <www.ndu.edu/WMDC Center>. 07 February 2014.
14. Rózsa, L. (2009) The motivation for biological aggression is an inherent and common aspect of the human behavioural repertoire. *Medical Hypotheses*, **72**, 217–219.
15. See for example, the studies described by Alibek in the Congressional Hearings on *Engineering Bio-Terror Agents: Lessons from the Offensive U.S. and Russian Biological Weapons Programs*. Subcommittee on Prevention of Nuclear and Biological Attack and Committee on Homeland Security, House of Representatives, One Hundred Ninth Congress, First Session, 13 July, 2005.
16. National Academies (2014) *Emerging and Readily Available Technologies and National Security: A Framework for Addressing Ethical, Legal, and Societal Issues*. Washington, D.C: National Academies Press, (p. ix).
17. Meselson, M. (2000) Averting the hostile exploitation of biotechnology. *The CBW Conventions Bulletin*, June, 16–19.
18. Reference 16, p. x.
19. ibid, p. 2–19.
20. ibid, p. 2–22.
21. ibid, p. 2–25.
22. ibid, p. 3–1.
23. ibid, p. 3–2.
24. Vogelstein, R. J. (2013) Advancing information superiority through applied neuroscience. *John Hopkins Apl Technical Digest*, **23** (4), 325–332.
25. Dando, M. R. (2014) Neuroscience advances and future warfare. Chapter 139 in J. Clausen and N. Levy (Eds), *Handbook of Neuroethics*. Dordrecht: Springer.
26. Development, Concept and Doctrine Centre (2011) Moral, legal and ethical issues. Chapter 5, pp. 5–1 to 5–12 in *The UK Approach to Unmanned Aircraft Systems*. Joint Doctrine Note 2/11, Ministry of Defence, London.
27. Marchant, G. E., Allenby, B., Arkin, R., Barrett, E. T., Borenstein, L. M., Gaudet, O. K., Lin, P., Lucas, G. R., O'Meara, R., and Silberman, J. (2014) International governance of autonomous military robots. *The Columbia Science and Technology Review*, **XII**, 272–314.
28. Reference 16, p. 3–21.
29. Dando, M. R. (1996) *A New Form of Warfare: The Rise of Non-Lethal Weapons*. London: Brassey's, (pp. 9–28).
30. Reference 16, p. 3–22.
31. Reference 29, pp. 116–135.
32. Kelle, A., Nixdorff, K. and Dando, M. R. (2006) *Controlling Biochemical Weapons: Adapting Multilateral Arms Control for the 21st Century*. Basingstoke: Palgrave Macmillan, (pp. 91–115).

33. Dando, M. R. (2007) Scientific outlook for the development of incapacitants. Chapter 6, pp. 123–148 in A. M. Pearson, M. I. Chevrier and M. Wheelis (Eds), *Incapacitating Biochemical Weapons: Promise or Peril*. Lanham: Lexington Books.
34. Kelle, A., Nixdorff, K. and Dando, M. R. (2012) *Preventing a Biochemical Arms Race*. Stanford: Stanford University Press, (pp. 61–87).
35. Lakoski, J. M., Murray, W. B. and Kenny, J. M. (2000) *The Advantages and Limitations of Calmatives for Use as a Non-Lethal Technique*. College of Medicine, Pennsylvania State University.
36. ibid, p. 7.
37. ibid, p. 6.
38. Committee on Military Intelligence Methodology For Emergent Neurophysiological and Cognitive/Neural Science Research in the Next Two Decades (2008) *Emerging Cognitive Neuroscience and Related Technologies*. Washington D.C: National Academies Press, (p. 1).
39. ibid, p. 2.
40. ibid, pp. 127–128.
41. ibid, p. 131.
42. ibid, p. 134.
43. Green, C. (2008) The potential impact of neuroscience research is greater than previously thought. *Bulletin of the Atomic Scientists: Roundtable*, 9 July.
44. Green, C. (2008) The role of NGOs in addressing concerns about neuroscience. *Bulletin of the Atomic Scientists: Roundtable*, 29 October.
45. Reference 38, pp. 134–135.
46. Committee on Opportunities in Neuroscience for Future Army Applications (2009) *Opportunities in Neuroscience for Future Army Applications*. Washington D.C: National Academies Press.
47. ibid, p. vii.
48. ibid, p. 5.
49. Committee on Assessing Foreign Technology Development in Human Performance Modification (2012) *Human Performance Modification: Review of Worldwide Research with a View to the Future*. Washington D.C: National Academies Press.
50. ibid, p. 1.
51. ibid, p. 3.
52. Kurihara, K. and Tsukada, K. (2012) Speechjammer: A system utilizing artificial speech disturbance with delayed auditory feedback. CoRR.abs/1202.6106. Available at http://orxiv.org/abs/1202.6106>. 10 October 2013.
53. Republic of Croatia (1999) *Evaluation of Biological Agents and Toxins*. BWC/AD HOC GROUP/WP.356/Rev.1, United Nations, Geneva, 19 July.
54. Kagan, E. (2001) Bioregulators as instruments of terror. *Clinics in Laboratory Medicine*, **21** (3), 607–618.
55. Aas, P. (2003) The threat of mid-spectrum chemical agents. *Prehospital and Disaster Medicine*, **18** (4), 306–312.
56. Kirby, R. (2006) Paradise lost: The psycho agents. *The CBW Conventions Bulletin*, **71**, 1–5.
57. Sutherland, R. G. (2008) *Chemical and Biological Non-Lethal Weapons: Political and Technical Aspects*. SIPRI Policy Paper No. 23, Stockholm International Peace Research Institute, Stockholm, November.

58. Royal Society (2012) *Brain Waves Module 3: Neuroscience, Conflict and Security*. Royal Society, London, February.
59. Mogl, S. (Ed.) (2012) *Technical Workshop on Incapacitating Chemical Agents*. Spiez Laboratory, Switzerland, February.
60. Regis, E. (1999) *The Biology of Doom: The History of America's Secret Germ Warfare Project*, p. 206. New York: Henry Holt.
61. Special Assistant to the Under Secretary of Defense (Personnel and Readiness) for Gulf War Illnesses, Medical Readiness and Military Deployments (2002) *Fact Sheet: Project Shipboard Hazard and Defense (SHAD)*. Department of Defense, Washington, D.C.
62. Dando, M. R. (2001) *The New Biological Weapons: Threat, Proliferation and Control*. Boulder: Lynne Rienner. (p. 62).
63. Ulrich, R. G., Sidell, S., Taylor, T. J., Wilhelmsen, C., and Franz, D. R. (1997) Staphylococcal enterotoxin B and related pyrogenic toxins. Pp. 621–630 in F. R. Sidell, E. T. Takafuji and D. R. Franz (Eds.) *Medical Aspects of Chemical and Biological Warfare*. Office of the Surgeon General, U.S. Army, Washington, D.C.
64. Reference 58, p. 46.
65. Dahan, A., Aarts, L. and Smith, T. W. (2010) Incidence, reversal, and prevention of opioid-induced respiratory depression. *Anesthesiology*, **112**, 226–238.
66. George, A. V. *et al*. (2010) Carfentanil – an ultra potent opioid.*The American Journal of Emergency Medicine*, **28** (4), 530–532.
67. Wax, P. M., Becker, C. E. and Curry, S. C. (2003) Unexpected 'gas' casualties in Moscow: A medical toxicology perspective. *Annals of Emergency Medicine*, **41** (5), 700–705.
68. Riches, J. R., Reid, R. W., Black, R. M., Cooper, N. J., and Timperly, C. M. (2012) Analysis of clothing and urine from Moscow theatre siege casualties reveals Carfentanil and Remifentanil use. *Journal of Analytical Toxicology*, **36**, 647–656.
69. Reference 65, p. 232.
70. ibid, p. 235.

7
Implications of Advances in Neuroscience

Introduction

Professor Matthew Meselson began his landmark 2000 article (Chapter 3) on 'Averting the Hostile Exploitation of Biotechnology' by suggesting that all major technologies have been exploited intensively for hostile purposes in the past and then he asked the question of concern here, 'Must this also happen with biotechnology, certain to be a dominant technology of the twenty-first century?' Significantly, Meselson then went on to quote directly from an award-winning essay by a military officer that had appeared in the summer 1989 volume of the US *Naval War College Review*. This essay assumed that biotechnology would be exploited intensively for hostile purposes and therefore stated that:[1]

> The outlook for biological weapons is grimly interesting. Weaponeers have only just begun to explore the potential of the biotechnology revolution. *It is sobering to realise that far more development lies ahead than behind.* [emphasis added]

Meselson comments that if this assumption is correct then biotechnology applications will significantly change the nature of weaponry and the conflicts within which such novel weapons are used.

So far in this book we have kept close to the here and now and avoided too much speculation about what could happen later in the century as the revolution in our understanding of life processes continues apace. However, it would be remiss not to give some consideration to possible longer-term developments before we consider the medium term possibilities and what might need to be done to minimise their hostile applications.

Jonathan Tucker attempted such an exercise in one of his last essays for *The Bulletin of the Atomic Scientists*.[2] Taking up Einstein's idea of using 'thought experiments' in his work on relativity, Tucker speculated on how there might be a place for expanding the scope of the definitions in the Chemical Weapons Convention to take account of emerging technologies. In one such thought experiment Tucker considered our developing knowledge of bioregulatory peptides such as Substance P, oxytocin and cholecystokinin. He argued that, whilst these serve as messenger substances in the nervous system in normal concentrations, 'at higher doses, [they] can have incapacitating or calmative effects'. Moreover, in his view, whilst natural peptides tend to be unstable in aerosols and would be degraded by enzymes in the body, 'it may be possible to develop structurally modified peptide analogues that are more persistent and capable of entering the brain from the bloodstream'. This led him to argue for such compounds to be included in the treaty's verification system and, more generally, from all of his thought experiments, to argue that the treaty needed to evolve in order to respond to changing technologies and not to stick blindly to the original negotiators' text if that was to become increasingly technically obsolete.

Lessons from parasitology

It is against that background that the question may be asked about where will our knowledge of the nervous system take us over the coming two, three or four decades up to the middle of the twenty-first century? One way to approach such a question is to ask what nature has evolved in regard to the malign manipulation of behaviour, and the place to look for that is in the field of parasitology. As one recent review noted:[3]

> Parasites often alter the behavior of their hosts in ways that are ultimately beneficial to the parasite or its offspring. Although the alteration of host behavior by parasites is a widespread phenomenon, the underlying neuronal mechanisms are only beginning to be understood.

So we know that the evolution of the interactions between parasites and the hosts they prey on has produced numerous examples of specific modifications of host behaviour – some very complex, as we shall see – and we are beginning to understand how this is achieved by modification of the host's nervous system.

The review concentrates on examples of direct modification of the host's central nervous system and gives a range of examples to illustrate the diversity of the possibilities. For example, an ant that becomes a victim of a parasitic fungus of the genus *Cordyceps* has its behaviour dramatically changed to benefit of the fungus. The spores of the fungus attach to the ant's cuticle and then germinate and get into the body via the tracheae (through which the ant breathes). The fungus then grows by feeding on the ant's body (avoiding the vital organs) and produces a chemical – as yet unknown – that causes the ant to climb to the top of a plant or tree and clamp itself in place there by holding on with its mandibles. When the fungus is ready to produce new spores and thus restart the cycle it kills the ant by eating its brain and the fruiting bodies of the fungus sprout out of the ant's body and shower the surrounding area with new infective spores.

An even more complex example is in the manipulation of spiders' behaviour by the parasitic wasp *Hymenoepimecis*. The wasp stings the spider host on its web to paralyse it temporarily so that the wasp can lay its egg on the spider. After recovering from the paralysis the spider appears to resume normal activity, two weeks later its behaviour changes significantly. The cause is not yet known, but as the review states,

> It appears that the [wasp] larva chemically manipulates the spider's nervous system to cause the execution of only one subroutine of the full orb web construction program while repressing all other routines.

The net result is that the web is now designed to support the wasp's cocoon so that it is suspended in the air. The wasp larva then eats the spider and pupates in the new, specifically constructed, web in order to restart the cycle.

In these two examples the neuronal mechanisms are, as yet, unknown, but most of the review is devoted to a discussion of an example where considerable progress has been made in recent years in understanding how the parasite achieves its objectives. This example again concerns the activity of a parasitic wasp (technically termed a parasitoid wasp as it kills its host in the process), the jewel wasp *Ampulex compressa*, and its usual victim, the cockroach *Periplaneta americana*.[4] The cockroach's central nervous system consists of a supra-oesophageal ganglion at the top of its head connected by two circumoesophageal connectives to a sub-oesophageal ganglion and then on to a series of ganglia in the thorax of its main body.[5] These ganglia of the thorax directly control the motor circuits for walking, but are under the control of the higher centres in the head.

The hunting wasp's first action is to sting the cockroach in its prothorax, which produces a brief paralysis of the forelegs. This allows the wasp to carry out a more difficult second sting into the cockroach's neck that puts venom into the two cerebral ganglia.[6] The second sting has two effects on the cockroach: first, when the initial leg paralysis ends it enters a period of self-grooming that lasts for about 30 minutes; second, the cockroach enters a prolonged hypokinetic state in which it does not initiate spontaneous walking. The self-grooming phase may have the advantage to the wasp of clearing the cockroach of bacteria and parasites on its cuticle prior to the laying of the wasp's egg, but it clearly keeps the cockroach in place whilst the wasp seeks a suitable hole to place the cockroach and egg in for safety.

When the wasp returns it cuts the cockroach's head antennae, feeds off the extruded liquid and then pulls the victim to the selected burrow. Interestingly, the cockroach is able to follow the wasp by walking, so the smaller wasp does not have to drag the cockroach to the burrow by itself. When inside the burrow the wasp lays its egg on the cockroach and covers up the entrance. If the egg is removed, experimentally, the cockroach resumes normal activity after a few days. But if this is not done the larva emerges from the egg and gets inside the cockroach to feed. The venom also appears to lower the metabolic rate of the cockroach so that although it does not feed or drink it provides fresh food for the larva for a few days until the larva pupates, to then emerge and restart the cycle four weeks later.

The hypokinetic state induced by the second sting appears to be quite specific to the initiation and maintenance of walking behaviour as, with suitable stimulation, the cockroach can still fly, right itself and, to a lesser extent, swim. Certainly, studies of wasps loaded with radioactive material have shown that the venom is localised to the ganglia of the thorax and:[7]

> around the midline of the sub-esophageal ganglion...and...in the central part of the supra-esophageal ganglion.

Moreover, experiments on stung cockroaches show that they:

> endured voltages more than eightfold higher than un-stung cockroaches before escaping the electric foot shocks, indicating their elevated threshold for walking initiation as well as their basic ability to walk upon reaching this threshold.

In short, the neuronal circuit for walking in the ganglia of the thorax is intact in the stung cockroach, but its resting state is such that it cannot easily start or maintain this activity because it lacks the necessary input from the higher centres.

Whilst the specific circuits in the cockroach central nervous system that are affected by the venom are yet to be elucidated in detail:[8]

> the prime candidates are neuromodulatory interneurons and, in particular, monoaminergic interneurons which descend from the SEG [sub-esophageal ganglion] to thoracic motor centres and/or ascend from the SEG to the SupEG [supra-esophageal ganglion].

The authors go on to add that:

> With respect to motivation, the role of monoaminergic systems ... may have been conserved throughout evolution. In insects as in mammals, CNS dopamine (DA) and octopamine (OA, which is the invertebrate analogue of noradrenaline), have been shown to profoundly affect motivation, arousal and locomotion.

At the very least, as we have seen in previous chapters, neuroscientists expect to learn much from the study of neuronal circuits in such simpler systems that will have an application in mammalian systems.

Still, the question might well be asked why, if there are so many examples in nature of the malign manipulation by parasites of the host's behaviour, so little is yet known about how such effects are achieved? One reason, of course, is that parasitism is such a diverse subject that the hijacking of the host's nervous system is but one facet that could be examined. Another reason could be the sheer difficulty of raising and studying two (or maybe three if there is an intermediate host) species in the laboratory. But probably the main reason for the lack of progress is that, until recently, the people studying this phenomenon did not often see themselves as being in the same field of endeavour because the species involved were so diverse. However, that may be changing as neuroscience itself becomes more central to the life sciences and a new field of neuroparasitology develops. Certainly, a meeting of many of those involved in such studies was convened in 2012 and the resulting papers presented in a special edition of the *Journal of Experimental Biology* in 2013. As the journal's introduction to these papers points out, there could be at least three different ways in which the behaviour of the host is manipulated by the parasite.[9] Firstly, because the host mounts an immune response to the parasitic infection and the parasite has to avoid that immune response:

> It may be a small evolutionary step from manipulating the host's immune system to prevent destruction, to manipulating it to secrete modulators that lead to a change in behaviour.

Secondly, as we have seen, the parasite may not alter the host's behaviour indirectly through the immune system but act directly on the operation of the nervous system itself. Finally, the introduction argues that the action may be even more indirect than through the immune system:

> Parasites also specifically target the expression of neurofunctional genes by secreting second messengers that directly impact gene expression, such as…the alteration of moulting hormones in caterpillars by baculovirus.

Moreover, the parasite is likely to have evolved to employ more than one strategy to affect the host's behaviour rather than having to rely on a single specific targeting mechanism.

Following the introduction, this special issue has an editorial by the guest editors and then a total of 18 papers (Table 7.1). These are grouped into four sections: alterations of host behaviour; neuroimmunology; toxoplasmosis; and new approaches.

This editorial gives more reasons to help us understand why host manipulation by parasites has not been elucidated in detail yet. As the editors explain, the physiological processes normally operating in the host are often not well understood and thus the impact of the parasite's activities on the host's physiology can be difficult to determine. Furthermore, the host's general response to infection (sickness behaviour) and the host's manipulation by the parasite can be very similar and thus difficult to distinguish. More importantly here, as the editors note:[10]

> Finally, many cases of parasite manipulation occur in host-parasite systems in which the host is not a typical model for behavioural neuroscience research.

Thus the work is made more difficult because of the lack of background information on the physiology and behaviour of the species involved.

Table 7.1 Papers on neural parasitology

Section 1	Alteration of host behaviour Eight papers including: Parasites: evolution's neurobiologists Parasites' manipulation of host personality and behavioural syndromes
Section 2	Neuroimmunology Three papers including: Immune-neural connections: How the immune system's response to infectious agents influences behaviour
Section 3	Toxoplasmosis Four papers including: *Toxoplasma gondii* infection, from predation to schizophrenia: can animal behaviour help us to understand human behaviour?
Section 4	New approaches Three papers including: Host-parasite molecular cross-talk during the manipulative process of a host by its parasite

Source: Modified from Adams, S. A. and Webster, J. P. (2013) Neural parasitology: How parasites manipulate host behaviour. *Journal of Experimental Biology*, **216**, 1–160.

Nevertheless, the editors are clearly optimistic about the possibilities of rapid progress being made in understanding some of these interactions. They point out also that many parasites have a predilection for targeting the nervous system because that protects them from the full force of the host's immunological defences as well as giving them direct access to the means of altering the host's behaviour. This is therefore important for understanding parasitology more generally. They also make clear that the problem of manipulation of host behaviour is not only a problem in regard to invertebrate hosts but also in regard to vertebrates, including human beings. Nowhere, perhaps, is this clearer than in the papers on toxoplasmosis in the third group of papers.

The papers in this group are, in part, concerned with the way that the parasite *Toxoplasma gondii* can invade the human brain and then alter the behaviour of its natural rodent intermediate host to enhance its transmission to its definitive feline host. But, as the editors note in their introduction to these papers, the parasite can:

> alters human behaviour and may be involved in the etiology of serious mental disorders such as schizophrenia.

So, if we want to get an idea of what may become possible later in this century, behavioural manipulation of vertebrate behaviour by the toxoplasma parasite would seem to be a good place to start.

Toxoplasmosis

Toxoplasma gondii was discovered in a rodent in the early years of the twentieth century but only became of interest to the medical community very gradually: first because of its damaging effects on the retina; then to obstetricians because of the risk of congenital toxoplasmosis; and more recently because of the risk of encephalitis amongst immunocompromised individuals. Only in the last 20 years, however, has the association between latent toxoplasmosis and various psychiatric and neurological problems become more and more obvious.[11]

T. gondii is a protozoan which can infect warm-blooded animals all around the world. However, it can only undergo full reproduction in its definitive hosts, members of the cat family. In these species it is able to reproduce sexually in the walls of the intestine and oocysts are shed in the cat's faeces. The problem for the parasite therefore is how to get back inside its definitive host. The problem is solved by the fact that many species, such as rodents, birds and humans, can become infected, for example by the ingestion of food contaminated by oocysts or by ingesting cysts embedded in raw or uncooked meat. Once inside a secondary host the parasite undergoes asexual reproduction and encysts in various tissues including the brain. Transmission back to the definitive feline host occurs when a cat eats something like an infected mouse or rat. Clearly, in evolutionary terms, there must be strong selective pressure on the parasite to evolve a means of making such transmission as effective as possible.[12]

Obviously, it is not in a rat's interest to spend time in places frequented by cats. So they have a strong aversion to places where cat odour can be detected, that is, as long as the rats are not infected by *T. gondii*. Infected rats, on the other hand, seem to have a fatal attraction to cat odour (urine), which appears to be quite specific and not caused, for example, by the odour of rabbits or even that of predatory mink. Such a change in behaviour in the rat would clearly be to the advantage of the parasite as it would assist its return to its definitive host by increasing the chance that the rat would be eaten by a cat.[13] Detailed studies of rodent behaviour when infected by the parasite have demonstrated that there is not just a change in the reaction to the potential predator's odour, but that there is also a range of other changes in behaviour that are likely to increase the risk of predation. For example, rodents infected by the parasite are more likely to be caught in traps than uninfected animals, suggesting that there are changes of a wider nature in the infected animal's exploratory behaviour.[14]

Detailed examination of the brains of infected mice led to the conclusion that there was not a well-targeted trophism towards any one region of the brain.[15] Rather, there was a scattering of parasite cysts with wide differences between individual infected mouse brains despite, as might be expected from the role of the amygdala in fear response,[16] there being a more consistent loading of cysts in this structure.

This evidence suggested that the parasite does not affect the host's behaviour towards the predator's odour through some direct physical disruption of the nervous system, and thus led scientists to the view that the disruption is probably caused by a more general disruption of chemical transmission in the brain. In particular, a raised level of dopamine in infected individuals has been found and two lines of evidence suggest that this disruption of dopamine transmission is caused by the parasite. First, the dopamine uptake inhibitor GBR12909 was found to modify the behaviour associated with the infection in rodents. Second, the parasite produces a key enzyme in the natural synthesis of dopamine.

The pathway leading from tyrosine is well known. First, the action of tyrosine hydroxylase metabolism produces L-DOPA and then decarboxylation of L-DOPA by another enzyme produces dopamine. It has been found that *T. gondii* produces a protein that is very similar to mammalian tyrosine hydroxylase – the rate-limiting enzyme in the production of dopamine. It has therefore been suggested that the high levels of dopamine found in the brains of infected mice could, at least in part, be the result of increased enzymatic activity related to the presence of the parasitic cysts.[17] Given the role of dopamine neurotransmission in decision-making and reward circuits in mammalian brains this is far from an unreasonable hypothesis.[18]

As we have seen, a parasite is unlikely to be able to rely on just one mechanism to achieve the goal of returning to its definitive host. Thus, there is good evidence that *T. gondii* is able to spread amongst its intermediate rodent hosts (and thereby have more chance of predation by cats) through infection during mating of rats. Again, it appears that this involves the manipulation of rodent behaviour, not of infected male rats but via the choice of mate by uninfected females.[19] Usually, a female rat would have mechanisms that lead her to avoid mating with parasitised males as this would be likely to reduce her, and her pups' survival chances. But males infected with this particular parasite seem to be more, not less, attractive to females than uninfected males, and this seems to be caused by the infected males producing higher levels of testosterone. So the infection has at least two routes to increase the chances of predation.

That, however, is unlikely to be the end of the story of the manipulation of host nervous, endocrine and immune systems. It has recently been shown, for example, that the manipulation of GABA and its receptors is probably one of the mechanisms by which *T. gondii* is able to migrate extensively from the gut to various host tissues, including the central nervous system. This involves hijacking the immune defence response. Dendritic cells (DC) of the immune system are ready to deal with invading pathogens and the mouse myeloid DC cells possess functional GABA receptors and GABA biosynthesis and secretion systems. After DC cells are infected with *T. gondii* they produce more GABA which acts on $GABA_A$ receptors of the DC cells and induces hypermobility. This mobility can be reduced by inhibition of GABA synthesis or by blockage of the receptors.[20] It is therefore likely that the parasite uses this mechanism to quickly establish itself widely following infection of the rodent.

So, even when mammalian hosts are involved it is clear that parasites have evolved means to manipulate behaviour in specific ways. This should not be a real surprise to us because we have all heard of the way in which the rabies virus changes the behaviour of dogs in quite dramatic ways. The fact that the chance and necessity of evolution and selective pressure have led parasites to develop multiple mechanisms should not blind us to the obvious conclusion that these involve elements of a quite specific manipulation of relevant neuronal circuits. But might it be possible that in the future specific aspects of human behaviour could be changed by specific manipulation of our complex nervous system? What indeed does the study of *T. gondii* infections have to tell us about that possibility?

When the parasite is shed from the cat rodents are obviously an ideal intermediate host to enable it to get back into a cat for further reproduction, but it can also infect many other species, including humans. It is thought that about a third of the human population, in developed countries, is infected. Over the last twenty years investigators have found that infected people do have differences, for example in reaction times (which might explain their increased likelihood of being in traffic accidents), and that, perhaps of most interest, infection seems to be related to schizophrenia and other mental illnesses.[21] The causes of such differences from uninfected people are yet to be elucidated, but it would be foolish to assume that these correlations have no specific cause, or that the causes will not be delineated in detail as neuroscience advances during this century. Moreover, it is not impossible that changes in neurotransmitters and hormones will be involved in the observed behavioural changes.[22]

Crucially, specific advances in neuroscience, for instance in understanding neurotransmitters, neuroreceptors and neuronal circuits, will be enhanced by the application of techniques in molecular biology derived from progress in genomics. Host manipulation by parasites presents a useful new way in which to understand the mechanistic basis of behaviour. As one researcher noted:[23]

> natural selection has acted on the genome of both the parasite and the host to control a single phenotype (behaviour in the host). Understanding diverse pathways from genes to phenotypes will help us tackle the important question of evolutionary biology: what is the mechanistic basis of animal behaviour?

Progress is certainly already being made through the use of such approaches, for example, by the study of the proteins produced in such interactions between parasite and host. As the reviewer also pointed out, an analysis of the proteome profiles of hairworms, which cause crickets to get into water (an unusual behaviour) in order that the parasitic worms can escape to mate:

> revealed a molecular cross-talk between the parasite inside the cricket's abdomen and the brain of the cricket. Specifically, the worm caused an upregulation of cricket Wnt proteins in the brain.

These proteins appear to act directly on the development of the brain, but it was also found that the worm:[24]

> secretes two families of proteins linked to the release of neurotransmitters ... and one family of proteins linked to the regulation of apoptosis [cell death] during the manipulative process.

It can confidently be expected that the use of molecular biology methodologies, like proteomic analysis, and increasingly powerful bioinformatics methods will further clarify how these parasites manipulate the behaviour of their hosts by interventions in the normal operations of host central nervous systems.

Conclusion

At the start of this chapter we asked whether, if the advances in neuroscience continue apace, it might be possible that over coming decades the capability to manipulate specific complex human behaviours might arise. As a means of investigating that issue we looked at current studies of the molecular arms race between parasites that manipulate the behaviour of their hosts and the hosts they infect.

It appears that the parasite is likely to have evolved a range of ways in which to achieve its objectives, but that within the whole attack there are specific manipulations of the host's central nervous system. Moreover, the growing capabilities, not just of neuroscience but also of molecular biology and bioinformatics, have been applied already to elucidate these interactions in detail. Finally, this process can be expected to continue and, in all probability, accelerate, as the field of neuroparasitism becomes more successful and attracts greater funding.

It hardly needs to be added that the manipulations discussed here are malign and are for the benefit of the parasite not the host. With this longer-term perspective in mind we can consider, in the next chapter, where malign (and purposeful) manipulation of the human nervous system is now, and where it might be in the medium term. The point is not that human beings with malign intent will try to replicate what parasites do but that, by using new knowledge and technologies, such human beings may to be able to devise and implement methods of malign manipulation much faster than in evolutionary processes.

References

1. Meselson, M. (2000) Averting the hostile exploitation of biotechnology. *The CBW Conventions Bulletin*, June, 16–19.
2. Tucker, J. (2011) Re-envisioning the Chemical Weapons Convention. *Bulletin of the Atomic Scientists*, Web Edition, 2 May. Available at http://www.thebulletin.org/features/re-envisioning-the-chemical-weapons-convention. 22 April 2015.
3. Libersat, F., Delago, A. and Gal, R. (2009) Manipulation of host behavior by parasitic insects and insect parasites. *Annual Review of Entomology*, **54**, 189–207.
4. ibid, pp. 197–203.
5. Gal, R, and Libersat, F. (2010) On predatory wasps and zombie cockroaches: Investigations of "free will" and spontaneous behavior in insects. *Communicative and Integrative Biology*, **3** (5), 458–461.
6. Gal, R., Kaiser, M., Haspel, G. and Libersat, F. (2014) Sensory arsenal on the stinger of the parasitoid jewel wasp and its possible role in identifying cockroach brains. *PLOS ONE*, **9** (2), 1–8.

7. Libersat, F. and Gal, R. (2013) What can parasitoid wasps teach us about decision-making in insects? *Journal of Experimental Biology*, **216**, 47–55.
8. ibid, pp. 52–53.
9. Knight, K. (2013) How pernicious parasites turn victims into zombies. *Journal of Experimental Biology*, **216**, i–iv.
10. Adams, S. A. and Webster, J. P. (2013) Neural parasitology: How parasites manipulate host behaviour. *Journal of Experimental Biology*, **216**, 1–2.
11. Flegr, J. (2013) How and why Toxoplasma makes us crazy. *Trends in Parasitology*, **29** (4), 156–163.
12. Webster, J. P., Kaushik, M., Bristow, G. C. and McConkey, G. A. (2013) *Toxoplasma gondii* infection, from predation to schizophrenia: Can animal behaviour help us understand human behaviour. *Journal of Experimental Biology*, **216**, 99–112.
13. ibid, p. 100.
14. Afonso, C. Paixâo, V. B. and Costa, R. M. (2012) Chronic Toxoplasma infection modifies the structure and risk of host behavior. *PLOS ONE*, **7** (3), 1–15.
15. Beremeiterová, M., Flegr, J., Kubērna, A. A. and Nēmec, P. C. (2011) The distribution of *Toxoplasma gondii* cysts in the brain of a mouse with latent toxoplasmosis: Implications for the behavioral manipulation hypothesis. *PLOS ONE*, **6** (2), 1–12.
16. Takahashi, L. K., Hubbard, D. T., Lee, I., Dar, Y. and Sipes, S. M. (2007) Predator odor-induced fear involves the basolateral and medial amygdala. *Behavioral Neuroscience*, **121** (1), 100–110.
17. Prandovszky, E., Gaskell, E., Martin, H., Dubey, J. P., Webster, J. P. and McConkey, G. A. (2011) The neurotropic parasite *Toxoplasma gondii* increases dopamine metabolism. *PLOS ONE*, **6** (9), 1–9.
18. House, P. K., Vyas, A. and Sapolsky, R. (2011) Predator cat odors activate sexual arousal pathways in brains of *Toxoplasma gondii* infected rats. *PLOS ONE*, **6** (8), 1–4.
19. Vyas, A. (2013) Parasite-augmented mate choice and reduction in innate fear in rats infected by *Toxoplasma gondii*. *Journal of Experimental Biology*, **216**, 120–126.
20. Fuks, J. M., Arrighi, R. B. G., Weider, J. M. Mendu, S. K., Jin, Z., Wallin, R. P. A., Rethi, B., Birnir, B., and Barragan, A. (2012) GABAergic signalling is linked to a hypermigratory phenotype in dendritic cells infected by *Toxoplasma gondii*. *PLOS ONE*, **8** (12), 1–16.
21. Flegr, J. (2013) Influence of latent *Toxoplasma* infection on human personality, physiology and morphology: Pros and cons of the *Toxoplasma*-human model in studying the manipulation hypothesis. *Journal of Experimental Biology*, **216**, 127–133.
22. Flegr, J. (2007) Effects of *Toxoplasma* on human behaviour. *Schizophrenia Bulletin*, **33** (3), 757–760.
23. Hughes, D. (2013) Pathways to understanding the extended phenotype of parasites in their hosts. *Journal of Experimental Biology*, **216**, 142–147.
24. Biron, D. G., Marche, L., Panton, F. Loxdale, H. D., Galeotti, N., Renault, L., Joly, C., and Thomas, F. (2005) Behavioural manipulation in a grasshopper harbouring hairworm: A proteomics approach. *Proceedings of Royal Society B*, **272**, 2117–2126.

8
The Search for Incapacitants

Introduction

In July 2014 the popular science journal *New Scientist* carried another of its many reports on advances in neuroscience research, this one concerning the mechanism that might produce human consciousness.[1] The article explained that towards the end of his life Francis Crick, the co-discoverer of the structure of DNA, and Christof Koch had speculated that the claustrum might be important in the production of consciousness.[2] The claustrum is a thin sheet-like structure of brain tissue located deep beneath the cortex which might integrate information from many different sources within the brain. The *New Scientist* was reporting on recent experiments carried out in the hope of helping a lady with epilepsy.

The scientists were placing an electrode in various parts of her brain and then briefly passing an electrical stimulation to the region where the electrode was placed. The report stated that the claustrum had never been stimulated in this way before and when it was something very unusual happened. The lady stared blankly and did not respond to the experimenters, but as soon as the stimulation stopped she resumed normal activity, although she had no memory of the period of stimulation. This happened repeatedly over a two-day period and the scientists were therefore able to eliminate a number of possible explanations for what was happening. Crucially, however, the lady was not being put to sleep but was still awake during the period of stimulation. Of course it is necessary to be careful in drawing conclusions from a single experiment, but this development suggested that further elucidation of the mechanisms that produce consciousness and other aspects of complex human behaviour are very likely to take place. Moreover, it might also encourage those who believe that precise chemical manipulation of behaviour will also become possible, despite well-founded doubts about any drug ever being free of side effects because when multiple targets are involved there will always be those who are affected.[3]

As the Royal Society report noted, currently weaponeers are interested in finding chemical agents with a rapid action and a short duration of effects and therefore:[4]

> contemporary interest has tended to focus on sedative-hypnotic agents that reduce alertness and, as the dose increases, produce sedation, sleep, anaesthesia and death.

There are many agents that could potentially be used to such effect,[5] but the Royal Society report focused particularly on opioids, benzodiazepines, alpha 2 adrenoreceptor agonists and various bioregulators, such as orexin.[6]

In Chapter 2 the alpha 2 adrenoreceptor agonist dexmedetomidine and its actions on the locus coeruleus (LC) noradrenaline neurons was given as an example of how a brain circuit might be affected by chemical agents. The neurons of the LC have widespread ramifications in the brain and produce noradrenaline that has a role in maintaining alertness and attention. However, the LC neurons also have receptors for noradrenaline and they function as alpha 2 adrenoreceptors that serve to inhibit the LC neurons in a negative feedback loop. Dexmedetomidine has the same effect on these receptors and can be used to reduce the output of noradrenaline from the LC and thus cause sedation. The Royal Society report notes that although it was originally developed as a veterinary drug dexmedetomidine has found a use in reducing the dose of primary anaesthetic agents used for human beings.

The action of fentanyl and its derivatives on the brain's natural receptors was also discussed in Chapter 6 in relation to the efforts by Russian special forces to break the Moscow theatre siege in 2002. The question then is, what progress is being made in the development of these and other potential incapacitating chemical agents mentioned in the Royal Society report, and are there other possible classes of such agents on the research horizon?

Current incapacitants

Our states of alertness and attention are clearly governed not only by specific neuronal circuits but also by our sleep/wake control system and broader circadian (day/night) rhythms. Yet, as we have seen in recent work on the claustrum, increasingly powerful techniques are enabling scientists to understand specific parts of these brain functions and how they are integrated to produce our behaviour. For example, we now know a great deal more about the complex nature of sleep than we did just fifty years ago.

What is of interest here, however, is how sleep is triggered and maintained, and how the awake state is produced and maintained. In short, it has become clear that:[7]

> Transitions from wake to sleep appear to be triggered by specific sleep-active ... neurons

and that:

> These neurons are tonically inhibited by wake-active ... neurons, and disinhibited after prolonged wakefulness.

So, in simple terms, there is a kind of flip-flop system in which sleep-active neurons are opposed to wake-active neurons and the two sets of neurons inhibit the other set's activity in a system of recriprocal inhibition (see Chapter 2). Clearly then, if more activity can be caused in one set, or if loss of activity can be caused in the other set, by chemical agents then the system can be pushed in a particular direction. This is just what happens, for instance, when dexmedetomidine is used to reduce the output of noradrenaline from the LC neurons and thus reduce the activity of this wakefulness-enhancing neuronal sub-system.

How does this link up with the Royal Society report's interest in benzodiazepines? The action of dexmedetomidine on the LC causes a decrease in its production of noradrenaline and, in part, this leads to GABA (gamma-aminobutryric acid) producing neurons being disinhibited so that they produce more GABA. This is the major inhibitory neurotransmitter in the brain and so the increase of GABA output causes an inhibition of other neurons and thus the sedative response.[8] As one major review noted:[9]

> A population of GABAergic neurons in the VLPO [ventrolateral preoptic area] show state-dependent firing patterns with the highest discharge rates during sleep ... The efferent projections of these neurons inhibit the centres promoting wakeful state.

One particular set of receptors – $GABA_A$ – is the target of benzodiazepines. These drugs do not directly affect the same part of the receptor as the natural transmitter, but, in what is known as an allosteric action, they interact with other parts of the receptor to enhance the impact of the natural transmitter. However, the $GABA_A$ receptor subtypes are very complex and it has proved difficult, so far, to be selective in activating one particular subtype and thus to elucidate its exact function in the brain.[10]

Yet, despite this, selective agents for the $GABA_A$ receptor are being developed:[11]

> Remimozolam (CNS 7056)…is a high-affinity and selective ligand for the BZD [benzodiazepine] site on the $GABA_AR$ [$GABA_A$ receptor].

This agent quickly induces sedation in human beings, is constructed so that it is metabolised rapidly in the body and therefore also has a short period of action,[12] and is undergoing clinical trials.[13] But it is important to recognise that CNS 7056 does not achieve this result by acting only on a specific subtype of the $GABA_A$ receptor but by modifiying the agent so that it is quickly rendered inactive by the action of the body's own enzymes. So even when considering benzodiazepines, drugs that have been subject to many years of research and development, the discovery of an agent with very specific effects on a particular circuit that underlies a behaviour of interest remains a distant goal.

Perhaps then a more interesting route for the weaponeer might be the consideration of novel neurotransmitter/receptor systems that are being discovered in the brain, and in that regard orexin and its receptors might be of great interest as disruption of this system produces the damaging sleep disorder called narcolepsy.

Orexin

Many of our basic physiological systems exhibit daily (circadian) rhythms. Most noticeably, we tend to sleep for about eight hours each night. During the latter half of the twentieth century we came to realise that sleep is not a passive state as recordings of brain activity showed very distinct phases, including rapid eye movement (REM), or paradoxical sleep, where the electrical recordings resemble those of the waking state. In this kind of sleep dreams seem to occur, and although eye movements take place voluntary muscle tone is lost. Despite such advances in our understanding of sleep, many people still suffer from debilitating sleep disorders such as insomnia, obstructive sleep apnoea and narcolepsy. Narcolepsy is characterised by excessive daytime sleep, catalepsy (sudden loss of muscle tone), dream-like experiences that occur just before sleep and sleep paralysis (inability to move as when falling asleep or just waking up).[14] The total amount of sleep and REM sleep stays in the same range as those not suffering narcolepsy, but the control mechanism seems to be disrupted as the person finds it difficult to maintain wakefulness and REM sleep seems to intrude into wakefulness, causing the loss of muscle tone and dream-like hallucinations. People suffering from this disorder lead very difficult lives but until recently we had little idea of what caused the problem.

Then, in 1998, as a recent review of the pharmacology and therapeutic opportunities to deal with narcolepsy noted:[15]

> two groups searching for new signaling molecules independently discovered the orexin neuropeptides and their receptors.

In less than a decade it also became clear that the effects of these neuropeptides were very important and, as the review stated:

> narcolepsy ... is caused by a loss of the orexin-producing neurons, and this has fueled a strong interest in developing orexin antagonists as a novel approach for promoting sleep and treating insomnia.

Of course, success in the search for such novel drugs is much to be desired to help people suffering from such a debilitating problem. Yet, elucidation of this mechanism of sleep disruption and the development of new sleep-inducing drugs might also open up pathways to new forms of incapacitation for those with less benign purposes in mind.

There are two orexin neuropeptides (Orexin A and Orexin B) and two orexin receptors (OX1R and OX2R). Orexin A has been shown to bind much more to OX1R than to OX2R, whereas Orexin B binds well to both receptors. Many studies have naturally been carried out to find specific information about these neuropeptides and their modes of action on these receptors and it is quite clear that they cause a depolarisation of neurons and a significant increase in the rate and length of activation. That is to say, these neurons have an excitatory effect on the neurons they innervate. Moreover, unlike attempts to modify behaviour through the monoamines (LC neurons) or GABA receptors (benzodiazapines), which can have significant side effects, the orexin system may be much more specific in its impact on the sleep/wake cycle and thus novel drugs may have few side effect problems, despite the likely implication of orexin in other central nervous system functions.[16] Certainly, major drug companies began to develop orexin antagonists very quickly after the initial discovery in the hope of helping people suffering from insomnia.[17]

More recent research has suggested that it is the OX2R receptor that may be predominantly involved in the control of the sleep/wake cycle and has thus led to the development of drugs targeted specifically at that receptor. For example, one study concluded that it had found that chemical JNJ1037049 specifically targeted OX2R and that:[18]

> Collectively, these findings highlight an essential contribution of the OX2R in modulating cortical activity and arousal, an effect that is consistent with the robust hypnotic effect exhibited by JNJ1037049.

The same paper reported on a different chemical, GSK1059865, that specifically targeted the OX1R receptor.

In such circumstances it is hardly surprising that a great deal of work is in progress by neuroscientists on the orexin system. Confusingly, the orexin neuropeptide is also called hypocretin by some researchers, with the receptors being called Hcrt-R1 and Hcrt-R2 as a result of names given by the two different investigations in 1998. However, it is quite clear that these hypocretin (orexin) neurons, in part, specifically target the monoamine-producing cells in the brain:[19]

> Hcrt neurons exhibit parallel firing patterns with monoaminergic neurons that represent tonic firing during wakefulness especially during active wakefulness, mild firing during slow wave sleep, and then silent during REM sleep, except in intensive firing at the transition to wakefulness.

The hypocretin (orexin) neurons innervate a number of monoaminergic systems within the brain and these probably have diverse functions when activated. What is particularly of interest here is the activation of the LC noradrenaline-producing neurons by the orexin neurons.[20]

This connection has been the subject of some detailed recent research. It should be noted that, although there are other noradrenaline-producing neurons in the brain that have diverse ramifications within other areas, only the cells of the LC, which produce some 50 per cent of the brain's noradrenaline output, innervate the cortex and thus directly affect the higher centres. Furthermore, recordings of awake animals show that the LC neurons fire tonically in the awake state, less during slow wave (normal) sleep and are almost silent during rapid eye movement sleep. LC neurons also fire phasically when the awake animal is presented with salient stimuli. Whilst the awake state is clearly supported by a variety of redundant mechanisms, as we have noted, cutting out the activity of the LC neurons has a pronounced sedative effect. More particularly, it has been shown that the noradrenaline input from the LC is crucial for maintaining the elevated membrane potential of neurons of the cortex in awake animals.[21]

To further investigate this system by specifically switching neurons of interest on or off some of the most advanced modern neuroscience tools have been used. Specifically, novel optogenetics[22] methods have been employed to examine the LC system and its links with the orexin (hypocretin) system in animal models. In this method genes for light-sensitive molecules are delivered to cells by viruses. However, specific promoters attached to the gene for the light-sensitive protein ensure that it is only manufactured in specific types of neurons. Then optical fibres are used to deliver pulses of light to either activate or silence the genetically targeted cells. So the functions of these cells can be examined in intact behaving animals. For example, if the orexin neurons are stimulated they normally activate the LC noradrenaline neurons and thus increase wakefulness, but if the orexin neurons are stimulated whilst the LC neurons are inhibited this blocks the transition from sleep to wakefulness. Additionally, it has been found that:[23]

> Acute optical activation of the hcrt neurons causes sleep-to-wake transitions over a time period of 10–30s, while stimulation of LC neurons causes sleep-to-wake transitions in less than 5s.

Such findings, naturally, suggest that the orexin neurons function to integrate numerous inputs related to sleep and wakefulness and then stimulate the LC (and other monoaminergic systems) to change behaviour outputs directly. Interestingly, from the perspective of this book, one of the funders for such research has been the United States Defense Advanced Research Projects Agency.[24,25]

There is much more to be discovered about the function of orexin and the causes of narcolepsy. One possibility that has received much attention is that narcolepsy could be an autoimmune disease caused by infection, or even by our attempts to vaccinate against infection.[26] The possibility that narcolepsy could be produced by artificial means in at least a proportion of a targeted population (assuming that some genetic susceptibility is involved) could be of considerable concern if taken up by those with malign intent. There is also much to be discovered, for example, about how the orexin system acts through the LC to affect the amygdala's role in threat learning.[27] Additionally, other techniques of modern genomics, such as the use of different animal models with well-known genomics, are likely to produce new insights.[28] Overall we can confidently expect further developments in our understanding of the mechanisms of sleep, wakefulness and arousal, and of drug targets and drug developments in the treatment of sleep disorders. It can also be confidently expected that some of these beneficial advances will carry potential dual-use implications.

It should also be noted that in recent years the problem of getting drugs into the brain past the blood brain barrier has been sidestepped by advances made in the intranasal delivery of numerous drugs.[29] There have, within this research area, been several successful studies of the intranasal delivery of orexins.[30] It seems likely that the pathways by which such chemical agents reach the brain from the nose are along the route of the olfactory nerve and that both intracellular and extracellular transport are possible, but with the extracellular pathway along the line of the nerve delivering agents in the greatest quantity and faster than other routes.[31] Most importantly, intranasal delivery to the brain works for large and large charged molecules that would not pass the blood-brain barrier in other circumstances. Therefore:[32]

> for a variety of growth factors, hormones, neuropeptides and therapeutics including insulin, oxytocin, orexin ... intranasal delivery is emerging as an efficient method of administration.

Again, this is good news for those seeking to get therapeutic agents into the brain, but it could have dual-use implications.

In the longer term developments in nanotechnology have also to be taken into account. There are certainly reasons to believe that nanoparticles could be used to enhance nose-to-brain delivery of drugs.[33] Nanotechnology describes a broad range of science and engineering that is designed to understand and manipulate material at the same kind of scale as molecular systems, such as cellular receptors that function in the brain. Nanotechnology developments have therefore raised concerns about how benignly intended developments might end up being used for malign purposes. Clearly, a standard agent might be encapsulated in a nanoparticle so that it was more stable in aerosol delivery, but it is also possible that nanoparticles could be designed to have a particular kind of cellular target once inside the body of the victim, or a nanoparticle could have a direct toxic effect itself.[34] A particular concern that has been noted by many authors is the possibility that nanoparticles carrying various bioregulators could be designed to pass through the blood-brain barrier and thus give direct access to weaponeers wishing to manipulate the operation of the central nervous system. Reference has been made to this approach being like a Trojan horse strategy.[35]

Yet, to many it may appear that such developments remain longer-term possibilities and therefore do not warrant too much attention at present. That would be to take a very optimistic perspective, particularly in regard to the possibilities of even benignly intended work giving rise to misperceptions and dangerous responses in kind that lead to further reciprocal responses.

Misperception processes

The history of the last century demonstrates how difficult it is to obtain accurate intelligence about chemical and biological weapons programmes.[36] For example, the UK and its western allies in the Second World War were very concerned that Germany might have an offensive biological weapons programme and therefore responded by developing biological weapons as a retaliatory capability. There was not an effective offensive biological weapons programme in Germany at the time, but there was an effective development of new nerve agents that the allies knew nothing about. Given that the States Parties to the Chemical Weapons Convention have not given a definitive ruling on the exact meaning of Article II.9(d) and the stated peaceful purpose of 'Law enforcement including domestic riot control purposes', there is a continuing danger that activities that could be associated with the development of novel incapacitants or responses to such new agents. These could be seen by other states as aspects of an offensive chemical weapons programme and thus bring the danger of erosion of the whole chemical and biological weapons non-proliferation regime as trust is progressively lost.

As we saw in Chapter 6 (Table 6.3) it is likely that a degradation market could produce a spiral of action and reaction, starting with the fear that incapacitants were being developed by states and thus leading to the response of producing countermeasures amongst other states. This would necessarily lead the responding state to investigate the science and technology of concern in some detail and may lead to the discovery of a novel agent. This concern could then lead the original state, even if innocent of any malign intent, to respond to the responder by continuing to investigate and develop new agents. Of course, other states could well follow such developments as best they can and consider whether they needed to investigate and develop countermeasures – with similar potential consequences becoming likely.

As incapacitating chemical agents have so far been closely related to drugs used in medical practice there is the possibility that work causing concern could be entirely benign and not at all related to any offensive or law enforcement purpose. However, there would be increasing difficulty in accepting that idea if the scientists involved appeared to be doing studies unrelated to medicine or said that they were indeed working on incapacitants. Such difficulties would be compounded if the scientists had links to, or worked for, defence organisations, particularly defence organisations with a longstanding programme related to chemical warfare agents, and where a state had previously developed or used an incapacitating chemical weapon. It seems likely that concerns that a law enforcement programme might actually be a cover for a new offensive programme would grow in such circumstances.

A detailed study of the literature recently illustrated how serious the possibilities of misperceptions of this kind might be.[37] Clearly, Russia must have had stocks of fentanyl derivatives and doctrine for, and training of, its special forces to have been able to break the 2002 Moscow theatre siege with the use of incapacitating agents.[38] Similarly, a Chinese company has advertised its BBQ-901 anaesthetic gun system, which is said to be effective up to 40 metres and on impact injects an incapacitating agent into the target. It is well known that the United States had long had an interest in the development of novel chemical non-lethal weapons,[39] until their representative at the Organization for the Prohibition of Chemical Weapons stated clearly in 2013 that the United States was not involved in the development of such chemical agents. The more detailed information available on the history of this work in the United States also shows how interest in such chemical agents can be switched from and between the police, intelligence and military and therefore provide many different development routes and groups whose activities in this area could cause concern in other countries. In India a defence scientist received an award in 2010 for his work on producing incapacitating chemical agents[40] and many of the major military powers in the world have had work carried out on such agents, which might produce concerns and responses.

Yet, it is not just major powers that could be seen as involved in the development of novel incapacitating chemical agents because the necessary scientific and technological capabilities are widespread. Thus, in Iran, work has recently been done on the aerosolisation of medetomidine, when it is standardly used in medicine by injection. This work could easily be subject to misperception by others since it was carried out in a university controlled by the military Revolutionary Guards.[41] Such studies are also evident in Europe where, in one country, there has been a programme, linked to the military, which has involved quite intensive studies of incapacitating chemical agents.[42] This work has involved the testing of aerosols of mixtures of agents, such as medetomidine and benzodiazepines and fentanyls, and despite indications that the work is no longer directly linked to the development of incapacitants,[43] the more recent work could still be misunderstood, even if it is for medical or veterinary purposes, because it involves many of the same people as the previous work.[44]

Conclusion

Clearly, there is every reason to expect that medical and scientific work will continue to explore the brain circuits underlying sleep, alertness and attention, and such work will certainly continue to lead to the development of new drugs to help people suffering from illnesses such as insomnia and narcolepsy. Moreover, we are unlikely to have come to the end of surprise discoveries related to these illnesses, such as the unexpected discovery and developments related to the role of orexin/hypocretin neuropeptides. But much of this work has potential dual-use implications and needs much more careful oversight than is presently available and, as we shall see in the next chapter, that is far from all that we need to worry about in the future.

References

1. Thomson, H. (2014) Consciousness – we hit its sweet spot. *New Scientist*, 5 July, 10–11.
2. Crick, F. C. and Koch, C. (2005) What is the function of the claustrum? *Philosophical Transactions of the Royal Society B*, **360**, 1271–1279.
3. Terbeck, S. and Chesterman, L. P. (2014) Will there ever be a drug with no or negligible side effects? Evidence from neuroscience. *Neuroethics*, **7**, 189–194.
4. Royal Society (2012) *Neuroscience, conflict and security*. Brain Waves Module 3, Royal Society, London. (p. 46).
5. For a detailed recent listing see Giordano, J. and Wiezman, R. (2011) Neurotechnologies as weapons in national intelligence and defence – an overview. *Synesis; A Journal of Science, Technology, Ethics, and Policy*, **2**, 138–154.

6. Reference 4, pp. 46–50.
7. Rosenwasser, A. M. (2009) Functional neuroanatomy of sleep and circadian rhythms. *Brain Research Reviews*, **61**, 281–306 (p. 286).
8. Nelson, L. E., Lu, J., Guo, T., Saper, C. B., Franks, N. P., and Maze, M. (2003) The α2-adrenoceptor agonist dexmedetomidine converges on an endogenous sleep-producing pathway to exert its sedative effects. *Anesthesiology*, **98**, 428–436 (p.428).
9. Saari, T. I., Uusi-Oukari, M., Ahonen, J. and Olkkola, K. T. (2011) Enhancement of GABAergic activity: Neuropharmacological effects of benzodiazepines and therapeutic use in anesthesiology. *Pharmacological Reviews*, **63**, 243–267 (p.250).
10. Olsen, R. W. and Sieghart, W. (2009) $GABA_A$ receptors: Subtypes provide diversity of function and pharmacology. *Neuropharmacology*, **56**, 141–148.
11. Reference 9, p. 254.
12. Kilpatrick, G. J., McIntyre, M. S., Cox, R. F., Stafford, J. A., Pacofsky, G. J., Lovell, G. G., Wiard, R. P., Feldman, P. L., Collins, H., Waszczak, B. L., and Tilbrook, G. S. (2007) CNS 7056: A novel ultra-short-acting benzodiazepine. *Anesthesiology*, **107**, 60–66.
13. *Remimazolam – anaesthetic/sedative*. See press releases by PAION AG at <info@paion.com>.
14. Kelle, A., Nixdorff, K. and Dando, M. R. (2006) *Controlling Biochemical Weapons: Adapting Multilateral Arms Control for the 21st Century*. Basingstoke: Palgrave Macmillan, (pp. 111–115).
15. Scammell, T. E. and Winrow, C. J. (2011) Orexin receptors: Pharmacology and therapeutic opportunities. *Annual Review of Pharmacology and Toxicology*, **31**, 243–268 (p. 144).
16. Hoyer, D. and Jacobson, L. H. (2013) Orexin in sleep, addiction and more: Is the perfect insomnia drug at hand? *Neuropeptides*, **47**, 477–488.
17. Parker, I. (2013) The big sleep: Looking for the next insomnia wonder drug. *The New Yorker*, 9 December, 50–63.
18. Gozzi, A., Turrini, G., Piccoli, L. *et al.* (2011) Functional magnetic resonance imaging reveals different neural substrates for the effects of Orexin-1 and Orexin-2 receptor antagonists. *PLOS ONE*, **6**, 1–12 (p. 1).
19. de Lecea, L. and Huerta, R. (2014) Hypocretin (orexin) regulation of sleep-to-wake transitions. *Frontiers in Pharmacology*, **5**, 1–7 (p. 2).
20. Carter, M. E., de Lecea, L. and Adamantidis, A. (2013) Functional wiring of hypocretin and LC-NE neurons: Implications for arousal. *Frontiers in Behavioral Neuroscience*, **7**, 1–7.
21. ibid, p.3.
22. Deisseroth, K. (2012) Controlling the brain with light. *Scientific American*, November, 49–54.
23. Reference 20, p. 5.
24. de Lecea, L., Carter, M. E. and Adamantidis, A. (2012) Shining light on wakefulness and arousal. *Biological Psychiatry*, **71**, 1046–1052 (see page 8 for acknowledgement of funders).
25. Reference 20, p. 5.
26. De la Harrán-Anta, A. K. and Garcia-Garcia, F. (2014) Narcolepsy as an immune-mediated disease. *Sleep Disorders*, **2014**, Article ID 792687. Available at <http://dx.doi.org/10.1155/2014/792687>. 18 July 2014.
27. Sears, R. M., Fink, A. E., Wigestrand, M. B., Farb, C. R., de Lecea, L., and LeDoux, J. E. (2013) Orexin/hypocretin system modulates amygdala-

dependent threat learning through the locus coeruleus. *PNAS*, **110** (50), 20260–20265.

28. Elbaz, I., Yelin-Bekerman, L., Nicenboim, J. *et al.* (2012) Genetic ablation of hypocretin neurons alters behavioral state transitions in zebrafish. *Neurobiology of Disease*, **32**, 12961–12972.
29. Chapman, C. D., Frey II, W. H., Craft, S. *et al.* (2013) Intranasal treatment of central nervous dysfunction in humans. *Pharm. Res*, **30**, 2475–2484.
30. Dhuria, S. A., Hanson, L. R. and Frey II, W. H. (2009) Intranasal drug targeting of hypocretin-1 (orexin-A) to the central nervous system. *Journal of Pharmaceutical Sciences*, **98**, 2501–2515.
31. Reference 29, p. 2476.
32. ibid, p. 2475.
33. Mistry, A., Stolnik, S. and Illum, L. (2009) Nanoparticles for direct nose-to-brain delivery of drugs. *International Journal of Pharmaceutics*, **379**, 146–157.
34. Kosal, M. E. (2014) Anticipating the biological proliferation threat of nanotechnology: Challenges for international arms control regimes, pp 160–174 in H. Nasu and R. McLaughlin (Eds) *New Technologies and the Law of Armed Conflict*. Berlin: Springer.
35. Committee on Military Intelligence Methodology for Emergent Neurophysiological and Cognitive/Neural Science Research in the Next Two Decades (2008) *Emerging Cognitive Neuroscience and Related Technologies*. Washington, D.C: National Academies Press. (see chapter 5 for a detailed discussion).
36. Dando, M. R., Pearson, G., Rozsa, L., Perry Robinson, J., and Wheelis, M. (2006) Analysis and Implications. Pp 355–373 in M. Wheelis, L. Rózsa and M. Dando (Eds), *Deadly Cultures*. Harvard: Harvard University Press.
37. Crowley, M. J. A. and Dando, M. R. (2014) *Down the Slippery Slope? A study of contemporary dual-use chemical and biological research potentially applicable to incapacitating chemical agent weapons*. Report Number 8, Biochemical Security 2030 Project, Bath, UK: University of Bath, October.
38. Riches, J. R., Reed, R. W., Black, R. M., Reid, R. W., Black, R. M., Cooper, N. J., and Timberly, C. M. (2012) Analysis of clothing and urine from Moscow theatre siege casualties reveals Carfentanil and Remifentanil use. *Journal of Analytical Toxicology*, **36**, 647–656.
39. Dando, M. R. (1996) *A New Form of Warfare: The Rise of Non-Lethal Weapons*. London: Brassey's.
40. *DRDO Newsletter*, volume 30, number 4, April 2010, Defence Research and Development Organisation, India.
41. Kamrampey, H. (2011) Aerosolisation of medetomidine hydrochloride as an incapacitating agent. *J. Passive Def. Sci. & Technol.*, **3**, 51–56.
42. Hess, L., Schreiberova, J. and Fusek, J. (2005) Pharmacological non-lethal weapons. *Proceedings of the 3rd European Symposium on Non-Lethal Weapons*. 10–12 May, Ettlingen, Germany.
43. Streda, L. and Patocka, J. (2014) Risk to the purpose and objectives of the Chemical Weapons Convention? *KONTAK*, February, 57–63.
44. Hess, L., Vatava, M., Schreiberova, J., Malek, J. and Horacek, M. (2010) Experience with a naphthymedetomidine-ketamine-hyaluronidase combination in inducing immobilization in anthropoid apes. *Journal of Medical Primatology*, **39**, 151–159.

9
Bioregulators and Toxins

Introduction

Although our sense of smell is important to us, as primates, our sense of the world is dominated by our visual system and we therefore lack an appreciation of how important smell can be to other animals. A rather dramatic demonstration of this was reported in early 2014. Much work on behaviour is carried out by experiments on rats and mice and experimenters can be men or women. Canadian researchers reported that exposure to men, but not to women, produced a robust stress response in these animals.[1] The effect occurred because rodents sensed male odours and even T-shirts that had been worn by men induced a similar response. An interesting question was raised by commentators: if the rodents were stressed by men but not by women what impact would that have had on decades of research in which nobody worried about the gender of the experimenters?

The potential impact of input to the olfactory system of some animals has been well understood since 1959 when, in a mammoth effort, Butenandt and his colleagues identified a moth sex pheromone called bombykol. They had to extract this chemical from a huge number of female moths, but were then able to show that it triggered male moths to fly towards the source of this chemical signal.[2] Since then we have come to understand a great deal more about how such pheromones function in the behaviour and interactions of many species, and more and more is being discovered as today's sophisticated chemical and biological techniques are applied to the study of such signaling chemicals. For example, it is no longer necessary to extract a potential pheromone from a huge number of animals because we can separate and identify minute quantities of such chemicals.

Most animals studied appear to have pheromones (species-wide chemical signals) so it might be expected that humans also have such signals. However, to date no such chemical has been identified.[3] Yet, a startling experiment with a two-person game showed that a dose of oxytocin delivered to the olfactory system of one player could dramatically affect their level of trust in the other player.[4] Oxytocin is important because it has long been studied and has many important bioregulatory functions in human physiology, including as a neuropeptide.[5] So, it is clear that if we want to consider the possibility of a malign manipulation of human beings through future advances in neuroscience we need to give some consideration to modern work on neuropeptides like oxytocin.

It will be recalled from Chapter 1, however, that as regards the the Biological and Toxin Weapons Convention, the World Health Organisation argued that such a bioregulator, if administered in unusual quantities, would be considered a toxin.[6] Thus bioregulators, if misused, come under the purview of the BTWC and the Chemical Weapons Convention. But there are many, many toxins in the natural world and these often target elements of the victim's nervous system. So we should begin the exploration of the malign potentialities by looking at natural toxins.

Toxins

The Australia Group (AG) is a multilateral organisation set up in the early 1990s, and much expanded since then, which seeks to restrict the threat of proliferation of chemical and biological weapons by operating an agreed control list for exports.[7] At the present time there are 19 toxins on the AG list and these are set out in Table 9.1. They include well-known toxins such as botulinum (except in forms for human administration and medical treatment), and ricin and saxitoxin, which are on a schedule list for inspection under the CWC. However, there are numerous toxins on the list that are not well known.[8]

States within the Australia Group will have had various good reasons for including these toxins on their list, but the list would also have been kept as short as possible because the longer the list the more difficult it is to set up effective export control measures. An insight into one probable major reason for inclusion of toxins on the AG List can be gained from a paper produced for a meeting of the Pugwash Scientists group just before the Second Review Conference of the Chemical Weapons Convention in 2007.[9] Volume II of the classic 1970s study, *The Problem*

Table 9.1 Toxins listed by the Australia Group

Toxins as follows and subunits thereof:[1]
1. Abrin
2. Aflatoxins
3. Botulinum toxins[2]
4. Cholera toxin
5. Clostridium perfringens alpha, beta 1, beta 2, epsilon and iota toxins
6. Conotoxin[2]
7. Diacetoxyscirpenol toxin
8. HT-2 toxin
9. Microcystin (cyanginosin)
10. Modeccin toxin
11. Ricin
12. Saxitoxin
13. Shiga toxin
14. Staphylococcus aureus enterotoxins, hemolysin alpha toxin, and toxic shock syndrome toxin (formerly known as Staphylococcus enterotoxin F)
15. T-2 toxin
16. Tetrodotoxin
17. Verotoxin and shiga-like ribosome inactivating proteins
18. Viscum Albut Lectin 1 (Viscumin)
19. Volkensin toxin

[1] Excluding immunotoxins

[2] Excluding botulinum toxins and conotoxins in product form that meet all of the following criteria:

- are pharmaceutical formulations designed for testing and human administration in the treatment of medical conditions;
- are pre-packaged for distribution as clinical or medical products; and
- are authorised by a state authority to be marketed as clinical or medical products.

Source: Modified from Australia Group (2014) *List of Human and Animal Pathogens and Toxins for Export Control*, January.

of Chemical and Biological Weapons – on novel chemical agents – took the tentative view that at that time the size and complexity of many such toxins made it unlikely that they would be considered as chemical weapons. The technology then was just not sufficiently well developed to be able to determine the chemical structure of many toxins accurately and to produce them by chemical synthesis. However, by 2007 it was clear that the required knowledge and technology was available to enable the production of a range of toxins (and bioregulators). As the authors noted:[10]

> Structural determination and cloning followed by expression have amplified production of protein toxins, especially now that some,

such as botulinum toxin and ricin, are known to comprise functional sub units.

What then do we know of these toxins today?

One good place to look for information is in textbooks of military medicine, and one major source of such textbooks is the Borden Institute. This is a United States Army Institute located at the Walter Reed Army Medical Center in Washington DC. It publishes volumes of the *Textbook of Military Medicine* free online.[11] Two of these volumes are of particular interest here: *Medical Aspects of Biological Warfare* and *Medical Aspects of Chemical Warfare*. Table 9.2 lists chapters of these two volumes relevant to this discussion.

Chapter 17 of the *Medical Aspects of Biological Warfare* deals with three groups of toxins: trichothecene mycotoxins from fungi, including T-2 toxins; marine algal toxins; and clostridial toxins from *Clostridium perfringens* (see Table 9.1). Chapters 14, 15, and 16 on staphlyococcal enterotoxin B and related toxins, ricin and botulinum toxin also cover toxins listed by the Australia Group.

Chapter 19 of the *Medical Aspects of Chemical Warfare* on Toxins: Established and Emergent Threats contains a long list of known toxins and their sources (in its Table 19.1), emphasising the diversity of the sources of toxins. The table covers many of the toxins listed by the AG, but does not cover a small number on the AG list. The emergent threats section of the chapter concentrates on palytoxin, which is not on the AG list, tetrodotoxin and saxitoxin, which are on the AG list, and brevetoxin which, again, is not on the AG list.

In an earlier version of the US Textbook of Military Medicine on *Medical Aspects of Chemical and Biological Warfare*,[12] the chapter Defence Against Toxin Weapons[13] gives us an understanding of the key aspects of

Table 9.2 Chapters on toxins

Medical aspects of biological warfare
Chapter 14. Staphylococcal enterotoxin B and related toxins
Chapter 15. Ricin
Chapter 16. Botulinum toxin
Chapter 17. Additional toxins of clinical concern
Medical aspects of chemical warfare
Chapter 19. Toxins: Established and emergent threats.

Source: Modified from *Medical Aspects of Biological Warfare* and *Medical Aspects of Chemical Warfare*. Available from the US Army Medical Department Center and School Portal.

this subject in relation to the focus of this book on neuroscience and the implications of advances in neuroscience. The chapter notes the diversity of the sources of toxins: bacteria; marine organisms; fungi; plants; and animal venom. However, it points out that there is not necessarily a correlation between the source of the toxin, or its molecular structure or size, and its mechanism of action. Crucially, it points out that there are just two general mechanisms of action: one is by damaging the membrane of the victim's cells; the other is by attacking the nervous system. As the chapter states,[14] 'neurotoxins exert direct effects on the nervous system functions'. Some toxins, for example staphylococcal enterotoxin B, also act indirectly on the nervous system, and thus behaviour, via primary effects on the immune system.

The threat from toxins is related to their ease of production, stability in storage and dispersal, and other standard issues related to weaponisation. Yet, it is obviously also crucially related to the lethality of the toxin itself. The chapter provides a list that compares the lethal dose of a range of toxins with those of classic chemical nerve agents.[15] Some examples are set out here in Table 9.3.

What is quite evident from this tabulation is the extreme lethality of neurotoxins such as botulinum toxin. Moreover, as production capabilities improve, other neurotoxins, such as saxitoxin and palytoxin, could become more of a threat. The AG list covers both militarily significant toxins, such as botulinum toxin, and agents such as ricin, which appears quite frequently in concerns about terrorist threats because it is relatively easy to prepare from castor oil seeds.[16] One point that needs to be understood is that natural toxins have resulted from evolutionary

Table 9.3 Comparative lethality of selected toxins and chemical agents

Agents	**LD50 (μg/kg)***
Botulinum toxin	0.001
Tetanus toxin	0.002
Clostridium perfringens toxins	0.1–5.0
Tetrodotoxin	8.0
Saxitoxin	10.0
VX	15.0
Soman (GD)	64.0
Sarin (GB)	100.0

* Dose required to kill 50% of those exposed calculated for laboratory mice.

Source: Modified from Sidell, F. R., Takafuji, E. T. and Franz, D. R. (Eds) (1997) *Medical Aspects of Chemical and Biological Warfare*. Office of the Surgeon General, Borden Institute, Washington, D.C., pp. 603–619.

processes between prey and predator and therefore resulting toxins are likely to be very difficult to make more effective. No such restriction is likely to apply to natural bioregulators and thus it is likely that modifications could make them much more effective and threatening if designed for malign purposes.

Bioregulators

Concerns about the possible use of toxins and bioregulatory midspectrum agents has a long history. Worries about increased capabilities for the production of toxins arose at the Second Five Year Review Conference of the Biological and Toxin Weapons Convention in 1986.[17] Surprisingly, as noted in Chapter 1, concerns about the possible misuse of bioregulators arose almost immediately in a document titled *Novel Toxins and Bioregulators: the Emerging Scientific and Technological Issues Relating to Verification of the Biological and Toxin Weapons Convention,* distributed to States Parties by the Canadian Government,[18] and in a contribution to the background paper on relevant scientific and technological advances, both for the 1991 Third Review Conference of the BTWC by the United States.[19] What should be noted is that these concerns were expressed before the genomics revolution facilitated the discovery of many more bioregulators during and after the 1990s.

These midspectrum agents were already the subject of proliferation concerns in the early 1990s as can be seen from another Canadian document of 1991 titled *Collateral Analysis and Verification of Biological and Toxin Research in Iraq,* which sought to examine research in Iraq from 1969 to 1991 through analysis of some 10,000 publications in the open literature.[20] This was done by searching for specific keywords in publications related to bioregulators, as set out in Table 9.4.

So, more than twenty years ago there were clearly numerous possible bioregulators of concern. As the Canadian study noted,[21] work on these

Table 9.4 Bioregulator key words

angiotensin	delta sleep-inducing peptide	gastrin	Substance P
atrial natriuretic peptide	dynorphin	gonadoliberin	thyroliberin
bombesin	endorphin	neurotensin	vasopressin
bradykinin	endothelin	neuropeptide Y	
cholecystokinin	enkephalin	somatostatin	

Source: Modified from Dando, M. R. (2001) *The New Biological Weapons: Threat, Proliferation and Control*. Boulder, CO: Lynne Rienner, p. 147.

topics did not necessarily imply any direct connection between the published work and a biological weapons programme but 'these key words are indicators that are considered to have a potential relationship to biological warfare research'.

As pointed out in Chapter 6, the reason for concern could sometimes be picked up in later publications. For example, in 1998 a paper on Substance P was published by scientists at the Defence Research Establishment, Division of Nuclear, Biological and Chemical Defence, in Sweden. The authors stated that:[22]

> The aim of the study was to determine the acute toxicity and effects on respiration of Substance P (SP), *a possible future warfare agent,* in guinea pigs when the substance was inhaled as an aerosol. [emphasis added]

The study demonstrated that when administered in this way, and protected by thiorphan against enzymatic breakdown, there were severe breathing problems in the experimental animals that probably resulted from bronchoconstriction caused by Substance P. The authors concluded that 'exposure to the substance at extremely low air concentrations may result in incapacitation in humans'. Whilst work on Substance P dates from the 1930s, it is only since the 1970s that it has been possible to show that it is but one example of a very large group of related bioregulators known to exist in many different animal species. It has also been possible to elucidate the structure of this peptide and demonstrate that it is the highly conserved terminal sequence of the, so-called, tachykinins that largely produces their biological actions.[23] It turns out that Substance P is not the only such tachykinin found in mammalian tissues. As a group these are called neurokinins and receptors for these bioregulators have also been well investigated.

Another example of an official level of concern was evident in a 2003 paper by Paul Aas of Norway's Defence Research Establishement.[24] The paper appeared in a special edition of *Prehospital and Disaster Medicine* and was titled 'The Threat of Midspectrum Chemical Warfare Agents'. Aas gave Substance P, and the related neurokinin A, ricin, saxitoxin and botulinum toxin, as examples of such midspectrum agents and argued that 'the primary target of these substances is the nervous system and many have a very high neurotoxicity'. Furthermore, Aas suggested that the use of such agents could be seen as advantageous for those with malign intent as they could be cost effective when used against concentrations of people or forces, and they could hamper military forces by making them don protective equipment or by causing panic amongst civilians. During the negotiations of the CWC in the early 1990s the UK certainly gave consideration to proposing that Substance P be included in the verification system schedules to serve as a marker for bioregulators, in the same way that saxitoxin and ricin are now intended to serve as markers for the different types of toxins.[25]

It could be argued that the use of Substance P as an agent for incapacitation would not really change the problem very much. Bronchoconstriction, after all, is still a rather simple method of incapacitation. This does not really signal a future in which chemical agents – bioregulators – are used to target much more complex behaviours. That may be a misreading of the problem that could arise in the future because bioregulators are also involved in complex behavioural regulation. This becomes clear if we take a closer look at one of the best-investigated bioregulators, oxytocin.

Oxytocin

Oxytocin and vasopressin (which has a very similar structure), function in the body as both hormones and neurotransmitters within the brain. These two bioregulators, which are produced in the brain, have been the subject of intensive study for over half a century. The amino-acid sequence of both were determined in the 1950s, the genes for both were cloned in the 1980s, and the structure of the oxytocin receptor was worked out in the early 1990s.[26] Each of these bioregulators has a structure of only nine peptides that differ in only two amino-acids. Yet these apparently simple molecules, and related molecules, have multiple complex roles in vertebrate species – including mammals like man. As the introduction to one major collection of essays noted, these two chemicals appear to be 'telling' the rest of the body what behavioural and physiological functions the brain wants carried out at any particular time.[27]

Whilst oxytocin and vasopressin are only found in mammals they are, in fact, members of a family of chemicals that are very ancient, and conserved in many vertebrate and invertebrate species. Indeed, in jawed vertebrates vasopressin-like and oxytocin-like neuropeptides derive from duplication of one ancestor gene. Functionally, these chemicals may have an original role in osmoregulation and fluid balance. Whilst the precise role of these chemicals may vary and cause different behaviours in different types of animals, they are often clearly involved in reproductive, parental and social behaviours. Fascinatingly, the venom cone snails use to catch prey contains conporessin-T, which is an analogue of vasopressin and may function as a vasopressin antagonist. In higher molluscs like octopus and cuttlefish there are two families of oxytocin-like and vasopressin-like peptides that have effects on learning and memory similar to those found in mammals.

What then do we know about the involvement of these neuropeptides in human brain and behaviour? A clue can be gained from the chapters in the last section of the book *Oxytocin, Vasopressin and Related Peptides in the Regulation of Behavior* that deal with studies of the role of these neuropeptides in human beings.[28] The chapters cover fear and anxiety, empathy and prosocial behaviour, sociability and social psychopathologies and autism, thus clearly indicating how important oxytocin (and vasopressin) are in the regulation of our behaviour. Another review paper on oxytocin that was concerned with pharmacotherapy argued that social behaviour is crucial for reproduction, protection against predators and brain development and that exclusion from the group is very dangerous for individuals.[29] Since behavioural pathologies such as autism can lead to exclusion, the authors argue that it is important to understand how they come about and how they might be treated. Thus the review went on to examine the progress in developing drugs related to oxytocin for that purpose. It is not necessary to go into the details of such advances here, but rather to understand how the capabilities to carry out such investigations have come about.

Historically, studies of the role of oxytocin in the mammalian brain were carried out on animals as such studies could not be carried out on human beings. However, the finding that intranasal administration of oxytocin led to it reaching the brain opened up the possibility of further studies of its function after delivery by aerosol. Moreover, developments in our understanding of the genetic variations in human oxytocin receptor genes allowed the impact of these variations to be better understood when oxytocin was delivered to the brain. Such studies, though, did not allow insights into the mechanisms underlying the impact of oxytocin on human behaviour and, given the known variations in these mechanisms in different mammalian species, this left explanations open to some doubt. The use of brain imaging techniques, however, has made it possible to begin to elucidate these mechanisms. Pharmacological fMRI (functional magnetic resonance imaging) is a technique which facilitates study of what happens in the brain when an agent like oxytocin is administered and allows, for example, its role in affecting processing of social signals, like fear, in the amygdala to be examined in considerable detail.[30]

It is not surprising that the military are interested in such developments. For example, the United States Defense Advanced Research Projects Agency (DARPA) put out solicitation SBI 32–001 in 2013 titled *Oxytocin: Improving measurement sensitivity and specificity*. The solicitation explained that:[31]

> Oxytocin is a hormone widely known for its role in reproduction and childbirth. More recently its role as a neuromodulator has been highlighted, particularly in facilitating pair bonding, maternal internal interactions, and trust behaviors.

That, as we have just seen, is a correct appreciation. The solicitation then continues:

> An explosion of research on the effects of oxytocin has ensued, and the hormone is listed in 213 ongoing or completed clinical trials on clinicaltrials.gov.

Furthermore, and of particular interest here:

> *Oxytocin also affects behaviors relevant to national security*. Oxytocin can impact behaviors ranging from whether two individuals trust each other, how someone reacts to stress, and even wound healing. [emphasis added]

Thus, the solicitation noted that understanding the function of oxytocin could be important across Department of Defense research programmes.

The text goes on to explain that, at present, research is difficult as there are different forms of oxytocin in the body and these could have different roles. Therefore an assay needs to be developed that can reliably distinguish between the different forms. The active form of oxytocin is cleaved from a somewhat larger prohormone. The role of this prohormone is as yet unclear, but it may be biologically active. The solicitation sets out the phases of research that are required to develop the assay and ends by describing some of the reasons why the military would be interested in the assay:

> Other military partners may be interested in this technology or a future derivative for the measurement of oxytocin *to understand influence*. [emphasis added]

It elaborates as follows:

> The DARPA program Narrative Networks is examining oxytocin in this context and would benefit from increased measurement specificity and sensitivity. Developments made under this SBIR effort could be transitioned, in synergy with funding from Narrative Networks, to provide better assays of oxytocin as it changes under narrative influence.

The DARPA Narrative Networks programme[32] aims, in part, to obtain an understanding of 'the effects narratives have on human psychology and its affiliated neurobiology'. Clearly, we have to cast the net broadly to understand military forces' interest in midspectrum agents. So, what are narrative networks all about?

The DARPA solicitation explains in this way:[33]

> Narratives exert a powerful influence on human thoughts and behavior. They consolidate memory, shape emotions, cue heuristics and biases in judgment, influence in-group/out-group distinctions, and may affect the fundamental contents of personal identity.

Therefore, in the view of DARPA;

> It comes as no surprise that because of these influences stories are important in security contexts: for example, they change the course of insurgencies, frame negotiations, play a role in political radicalization, influence methods and goals of social movements.

Therefore, understanding how stories can play such roles in conflict situations is considered an important objective by DARPA.

The solicitation asks for proposals in three separate, but linked, technical areas. Technical Area One is to ascertain:[34]

> exactly what function stories enact, and by what mechanisms they do so, is a necessity to effectively analyse the security phenomena shaped by stories.

Technical Area Three states:[35]

> In order to understand exactly how narratives influence human behavior, models must be developed that simulate these influences and directly measure their impact.

But it is their second technical area, that of narrative neurobiology, that is crucial here. The DARPA text states that:[36]

> Since the brain is the proximate cause of our actions, stories have a direct impact on neurobiological processes of both the senders and the receivers of narratives. Understanding how stories inform neurobiological processes is critical if we are to ascertain what effect stories have on the psychology and neurobiology of human choices and behaviors.

The text continues:

> The primary goal of Technical Area Two is to revolutionize our understanding of how narratives and stories influence our underlying neurobiology at multiple levels of analysis, ranging from basic neurochemistry, to the systems level, to big-picture systems-of-systems analysis.

Indeed, the text goes on to state that Technical Area Two 'serves as the neurobiological and neurochemical backbone of the narratives identified and analysed'. It is divided into five sub-goals, as set out in Table 9.5.

Sub-goal 1 is of particular interest here. The solicitation asks for proposals to:[37]

> Determine if narratives uniquely modulate human hormone or neurotransmitter production. Determine if production and uptake of behaviorally important neurotransmitters such as *oxytocin* and serotonin is influenced by narratives, and in what way. Identify *novel neurotransmitters* or *other biologically active molecules* modulated by narrative influence. [emphases added]

Table 9.5 Sub-goals of Technical Area Two

1. Assay narrative effects on our basic neurochemistry;
2. Understand narrative impact in the neurobiology of memory, learning and identity;
3. Assess narrative influence in the neurobiology of emotions;
4. Examine how narratives influence the neurobiology of moral judgement;
5. Determine how narratives modulate other brain mechanisms related to social cognition.

Source: Modified from Department of Defense (2013) 2013.2 SBIR Solicitation: *SB132–001: Oxytocin: Improving measurement sensitivity and specificity*. Available at <http://www.sbir.gov/print/401741>.

Clearly, DARPA is interested in the function of what the BTWC and CWC would see as relevant midspectrum agents.

This is not by any means a small low-cost programme that DARPA initiated in 2011. For example, Arizona State University was awarded Grant Number D12AP00074 for a total of $6,235,816 for work on a project titled 'Toward Narrative Disruptors and Inductors: Mapping the Narrative Comprehension Network and Persuasive Effects'.[38] Moreover, the work on the DARPA programme has involved very well-known individual scientists and research groups.[39] Whilst some might view this programme as a sensible security effort to help us understand fundamentalist radicalisation and how it might best be prevented, others might worry about what other purposes the knowledge gained might be put to in current circumstances.

Conclusion

There is a danger that BTWC States Parties will concentrate their attention on biological agents and CWC States Parties will concentrate their attention on chemical agents, and therefore that midspectrum toxins and bioregulators will consequently be neglected. Rather than there being a joint coverage of these agents there will instead be a gap between the coverage provided where the two conventions converge. It seems very important that such a gap is not allowed to develop, as there are numerous bioregulators and toxins and many ways in which advances in our knowledge of their functions could be relevant in maintaining the prohibitions embodied in the BTWC and CWC. Thus, the subject of the next chapter is the question of how well the conventions have coped with such scientific and technological change and may cope in the future.

References

1. Sorge, R. E., Martin, L. J. Isbester, K. A. *et al.* (2014) Olfactory exposure to males, including men, causes stress and related analgesia in rodents. *Nat Methods,* **11** (6), 629–632.
2. Wyatt, T. D. (2014) *Pheromones and Animal Behaviour: Chemical Signals and Signatures*, Second Edition. Cambridge: Cambridge University Press.
3. Reference 2, Chapter 13, On the scent of human attraction: human pheromones?, pp. 275–303.
4. Kelle, A., Nixdorff, K. and Dando, M. R. (2012) *Preventing a Biochemical Arms Race*. Stanford: Stanford University Press. (pp. 79–87).
5. Choleris, E., Pfaff, D. W. and Kavaliers, M. (2013) *Oxytocin, Vasopressin and Related Peptides in the Regulation of Animal Behavior*. Cambridge: Cambridge University Press.

6. World Health Organisation (2004) *Public Health Response to Biological and Chemical Weapons: WHO Guidance*, Second Edition. Geneva: WHO. (p. 216).
7. Mathews, R. J. (2004) The development of the Australia Group Export Control List of Biological Pathogens, Toxins and Dual-Use equipment. *The CBW Conventions Bulletin*, **66**, 1–4.
8. Australia Group (2014) *List of Human and Animal Pathogens and Toxins for Export Control*, January. Available at <http://www.australiagroup.net/**en**?/human_animal_pathogens.html>. 21 July 2014.
9. Phillips, A. P. and Robinson, J. P. P. (2007) *The CWC and Chemicals of Biological Origin*. Paper for the OPCW Academic Forum, The Hague, 18–19 September 2007 and Background Paper for the 27th Pugwash CBW Workshop, *Moving Forward After the Sixth BWC Review Conference*, Geneva, 8–9 December, 2007.
10. Reference 9, p. 3.
11. Available from the US Army Medical Department Center and School Portal. Borden Institute, Office of the Surgeon General, US Army.
12. Sidell, F. R., Takafuji, E. T. and Franz, D. R. (Eds) (1997) *Medical Aspects of Chemical and Biological Warfare*. Office of the Surgeon General, Borden Institute, Washington, D.C.
13. Franz, D. R. (1997) Defense against toxin weapons, pp. 603–619 in Sidell, F. R., Takafuji, E. T. and Franz, D. R. (Eds) *Medical Aspects of Chemical and Biological Warfare*. Office of the Surgeon General, Borden Institute, Washington, D.C.
14. Reference 13, p. 609.
15. Reference 13, p. 607.
16. See, for example, BBC News: US and Canada (2014) *Shannon Richardson gets 18-year jail term for posting ricin*, 17 July. Available at <http:// www.bbc.co.uk/world-us-canada-28339653>. 25 July 2014.
17. Dando, M. R. (1994) *Biological Warfare in the 21st Century: Biotechnology and the Proliferation of Biological Weapons*. London: Brassey's. (p. 74).
18. ibid, p. 138.
19. Dando, M. R. (2001) *The New Biological Weapons: Threat, Proliferation and Control*. Boulder: Lynne Rienner (pp. 37–38).
20. ibid, p. 147.
21. ibid, p. 146.
22. Koch, B., Edvinsson, A. A., and Koskinen, L-O. D. (1998) Inhalation of substance P and thiorphan: Acute toxicity and effects on respiration in conscious guinea pigs. *Journal of Applied Toxicology*, **19**, 19–23.
23. Reference 19, pp. 73–76.
24. Aas, P. (2003) The threat of midspectrum chemical warfare agents. *Prehospital and Disaster Medicine*, **18** (4), 306–312.
25. Walker, J. R. (2012) *The Leitenberg-Zilinskas History of the Soviet Bioweapons Program*. Harvard Sussex Program Occasional Paper No. 2, Harvard Sussex Program, University of Sussex
26. See reference 4, p. 82.
27. See reference 5, Preface pp. viii–ix.
28. Reference 5, Chapters 17–20.
29. Viero, C., Shibuya, I., Kitamura, N. *et al.* (2010) Oxytocin: Crossing the bridge between basic science and pharmacotherapy. CNS *Neuroscience and Therapeutics*, **16**, e138–e156.

30. Zink, C. F. and Meger-Lindenberg, M. (2012) Human neuroimaging of oxytocin and vasopressin in social cognition. *Hormones and Behavior*, **61** (3), 400–409.
31. Department of Defense (2013) 2013.2 SBIR Solicitation: *SB132–001: Oxytocin: Improving measurement sensitivity and specificity*. Available at <http://www.sbir.gov/print/401741>. 29 July 2014.
32. Defense Office (2011) *Broad Agency Announcement: Narrative Networks*, DARPA-BAA-12–13, 7 October, (p. 4).
33. ibid, p. 5.
34. ibid, p. 6.
35. ibid, p. 9.
36. ibid, p. 7.
37. ibid, p. 8.
38. Arizona State University (2013) *Narrative Networks (N2) – Phase I Progress, Status and Management Report*. Quarterly Progress Report 1 April 2013 through 30 June 2013. Arizona State University, 12 July.
39. See, for example, *Project: Neurobiology of Narrative Framing: A Collaboration Between the USC Institute for Creative Technologies and the USC Brain and Creativity Institute*. Available at <http://narrative.ict.usc.edu/neurobiology-of-narrative-framing.html>. 26 August 2014.

Part III
The Future

10
The BTWC and CWC Facing Scientific Change

Introduction

Both the BTWC and the CWC have built-in mechanisms designed to deal with scientific and technological change. Article XII of the BTWC states that:

> Five years after the entry into force of this Convention ... a conference of States Parties to the Convention shall be held at Geneva, Switzerland, to review the operation of the Convention, with a view to assuring that the purposes of the preamble and the provisions of the Convention ... are being realised.

Subsequently, this evolved into a system of regular five-yearly reviews in 1980, 1986, 1991, 1996, 2001–2002, 2006 and 2011. Significantly, Article XII ends by adding that 'such review shall take into account any new scientific and technological developments relevant to the Convention'.

The CWC also has a requirement for regular five-yearly reviews in paragraph 22 of Article VIII and again this states that 'such reviews shall take into account any relevant scientific and technological developments'. However, the CWC is a much more complex treaty and paragraph 21 of Article VIII also requires that the Organisation for the Prohibition of Chemical Weapons shall:

> Review scientific and technological developments that could effect the operation of this Convention and, in this context, directs the Director-General to establish a Scientific Advisory Board to enable him ... to render specialized advice in areas of science and technology relevant to this Convention, to the Conference, the Executive Council or States Parties.

Additionally, as the organisations representing chemists are unified in the International Union of Pure and Applied Chemistry (IUPAC) it has been possible for the Scientific Advisory Board (SAB) to call on IUPAC to hold a meeting prior to the three five-yearly reviews to help inform the Director-General's report to the Review Conference.

History of the BTWC

It is interesting to note that, following the failure of the 2001–2002 Fifth BTWC Review Conference to agree a Final Declaration, the United Kingdom produced a Green Paper on what might be done to strengthen the Convention and this argued, in part, for a more regular review of relevant science and technology, stating:[1]

> in view of the dramatic pace of technical change in the life sciences as described here, an open-ended body of government and non-government scientists should meet every one or two years to review the rate of change and to assess their implications for the convention and measures to strengthen it.

However, despite the fact that State Parties to such agreements have a wide range of possible mechanisms for assessing the impact of scientific and technological change open to them,[2] and that scientists had obviously been heavily involved in the previous discussions of Confidence Building Measures and the VEREX process (which examined potential verification measures), the UK's suggestion was not taken up. Therefore, the primary means of assessing the impact of science and technology on the Convention has been through the five-yearly review conferences.

At these review conferences a system developed whereby States Parties that wished to do so could contribute to a background paper on science and technology that was then circulated to all State Parties. Until recently scientific and technological changes were considered mostly in relation to the scope of the Convention set out in Article I, and it was not at all obvious that much consideration was given to the background paper during the review. Nevertheless, scientific and technological considerations did have an impact on the understandings reached in regard to Article I in the final documents of successive review conferences.

A joint contribution to the background paper by the three Depositary States (the UK, USA and the Soviet Union) for the First Review conference in 1980 was quite sanguine about the impact of scientific change and the *Final Declaration*[3] stated, in regard to Article I, that it 'has proved sufficiently comprehensive to have covered recent scientific and technological developments relevant to the convention'.

Yet, by the time of the Second Review Conference in 1986 the background paper reflected growing concerns about the capabilities in recombinant DNA technology, microencapsulation and large-scale toxin production, and these concerns impacted the *Final Declaration*,[4] which stated in regard to Article I that:

> The Conference, conscious of the apprehensions arising from relevant scientific and technological developments, inter alia, in the fields of microbiology, genetic engineering and biotechnology, and the possibilities of their use for purposes inconsistent with the objectives and the provisions of the Convention, reaffirms that the undertakings given by States Parties in Article I applies to all such developments.

And further that:

> The Conference reaffirms that the Convention unequivocally applies to all natural or artificially created microbial or other biological agents or toxins whatever their origin or method of production, consequently, toxins (both proteinaceous and non-proteinaceous) of a microbial, animal or vegetable nature and their synthetically produced analogues are covered.

Subsequent *Final Declarations* from later Review Conferences continued to elaborate understandings of different aspects of the life sciences that were within the scope of Article I before the Sixth Review Conference of 2006 concluded that 'the Conference reaffirms that Article I applies to all scientific and technological developments in the life sciences and other fields of science relevant to the Convention'.

It can be concluded then that the States Parties were able, at these Review Conferences, to monitor the increasing capabilities of the life sciences that were relevant to the Convention and to agree that these capabilities were covered by Article I. However, it is also clear that they were unable to decide on what might best be done to minimise the threat that these developments posed to the Convention. The agreement to hold meetings in the Intersessional Processes (ISPs) after the broken 2001–2002 Fifth Review Conference and the 2006 Sixth Review Conference did allow for more time to be devoted to discussions of some scientific subjects, such as codes of conduct and oversight systems for scientific research, but again the Review Conferences did not produce effective decisions and actions on what might best be done.

Given that diplomats have considerable flexibility to decide how they will structure and operate their meetings,[5] it is not surprising that there were numerous proposals put forward at the 2011 BTWC Seventh Review Conference for how the following ISP could be made more effective and efficient.[6] However, many of these hopes were not realised in the eventual modest outcome of the review.[7] One advance that was made was the decision to have three standing agenda items (SAIs) for the meetings at Expert and States Parties levels each year through to the Eighth Review Conference in 2016. Although decisions could not be taken until that Review Conference this did at least open up the possibility of substantial and cumulative discussions to be undertaken on these SAIs and, crucially, one of the topics selected for an SAI was concerned with science and technology.

The briefest glance at the topics set for the SAI[8] on science and technology (Table 10.1) immediately raises the question of how a substantive and cumulative outcome could be possible given that both the Expert-level and States Party-level meetings were only allocated one week in Geneva each year and that there are many other issues that have to be on the agenda.

It is important to note here that issues related to the possible misuse of neuroscience had been prominent in a number of the State Party contributions to the background paper on relevant science and technology for the Seventh Review Conference. The UK's contribution, for example, had a specific section on neuroscience which included:[9]

> 120. In this field, technologies for the discovery and development of compounds that act on the central nervous system (CNS) are of particular relevance to the BTWC. These include advances in molecular and genetic neuroscience, drug discovery technology, bioregulators and drug delivery to the brain.

The UK took up the issue of neuroscience again in a Working Paper for the July 2012 Meeting of Experts. In this paper the UK suggested that neuroscience developments had to be included in the work of the ISP through to 2016:[10]

> Although neuroscience is not specifically mentioned in the list of topical scientific subjects to be addressed by the new intersessional process, advances in production, dispersal and delivery technologies

Table 10.1 Standing agenda items on science and technology

Review of developments in the field of science and technology related to the Convention
22. The Conference decides that the following topics will be addressed under the Standing Agenda Item on review of developments in the field of science and technology related to the Convention: (a) new science and technology developments that have potential for uses contrary to the provisions of the Convention; ... (d) voluntary codes of conduct and other measures to encourage responsible conduct by scientists, academia and industry; (e) education and awareness-raising about risks and benefits of life sciences and biotechnology; ... (g) any other science and technology developments of relevance to the Convention.
23. The following topical scientific subjects will be considered in the years indicated: (a) advances in enabling technologies, including high-throughput systems for sequencing, synthesising and analysing DNA; bioinformatics and computational tools; and systems biology (to be considered in 2013); ... (d) advances in production, dispersal and delivery technologies of biological agents and toxins (to be considered in 2015).

Source: Table created by the author, source of data from, Seventh Review Conference of the States Parties to the Convention on the Prohibition of the Development, Production and Stockpiling of Bacteriological (Biological) and Toxin Weapons and on Their Destruction (2012) *Final Declaration*. BWC/CONF.VII/7, United Nations, Geneva, Switzerland, 13 January (pp. 23–24).

> of biological agents and toxins is to be considered in 2015 ... the UK calls upon States Parties to come prepared at that meeting.

The topic to be addressed in 2015 can be seen in item (d) of paragraph 23 in Table 10.1. The Working Paper continued to press the importance of neuroscience as follows:

> In the meantime the UK favours placing the implications of neuroscience for the BTWC (and the CWC) as part of the regular scientific and technological review discussions in the intersessional process.

That is to say that neuroscience should be added as an eighth topic to the seven already listed in points (a) to (g) in paragraph 22 as shown in Table 10.1.

The Meeting of Experts in 2012 was, perhaps, not a fair test of the new system as it was held soon after the Review Conference and with a change in the designated Chairman, but with only six hours devoted to the science and technology SAI it is hard to see how proper attention could be given to such complex technical issues. At the Meeting of States Parties later in the year the science and technology SAI received two fewer hours – being allocated just four hours to cover the items set out for regular review and the special topic for the year (as shown in Table 10.1).

In fact the most interesting part of the later Meeting of States Parties was probably in relation to a Working Paper by South Africa titled *The intersessional process: comments and proposals*. In regard to the SAI on relevant science and technology this stated that in the Meeting of Experts:[11]

> 5. Some excellent presentations were given by experts in which the very complicated issues were explained in simple terms. However, there was no substantive engagement on these presentations and therefore, opportunities to come to useful common understandings were lost. A number of very useful discussions took place during lunchtime side events, but these were not attended by all delegations or part of the formal MXP [Meeting of Experts].

A report[12] of the Meeting of States Parties noted that following a presentation of this Working Paper by South Africa 'there was a lively discussion with some 19 States Parties making statements and interventions in the subsequent hour'.

The problem of lack of action by States Parties to the BTWC was well illustrated in relation to the topic of awareness-raising and education of life scientists (see point (e) of paragraph 22 in Table 10.1) in the run-up to the Meeting of States Parties in 2012. What needed to be done had been well analysed during the previous ISP in the 2008 meetings and 11 States Parties had reported their individual efforts in an important Working Paper for the Seventh Review Conference.[13] However, a careful comparison with the work of the IAEA (International Atomic Energy Agency) in developing an International Nuclear Security Education Network (INSEN) showed just how far behind the States Parties to the BTWC are in tackling the security education of relevant scientists.[14] A clear need was for a more effective and efficient ISP to be put in place so that critical decisions could be agreed in later years of the ISP.

Presumably in response to concerns expressed in 2012 the chair of the 2013 meetings of the BTWC States Parties, Ambassador Körömi of

Hungary, sent a formal letter[15] to all States Parties in February 2013 explaining how she hoped, without moving outside the mandate set by the Seventh Review Conference, to 'improve the process of preparing for our meetings, make more efficient and productive use of our precious meeting time, and enhance the practical value of the programme'. So the task for 2013 was clearly set out by the Chair but the question was, could States Parties meet the challenge?

History of the CWC

The Chemical Weapons Convention is a more recent and much more complex agreement than the Biological and Toxin Weapons Convention. The CWC also has a major international organisation, the OPCW, fully capable of implementing decisions taken by States Parties, as it has amply demonstrated in the verified destruction of most of the huge lethal chemical weapons stocks that had been built up during the last century. However, now it has to turn its attention to non-proliferation – the prevention of the re-emergence of chemical weapons – a task similar to that of States Parties to the BTWC in preventing the re-emergence of biological weapons. The question is, can States Parties to the CWC and the OPCW take and implement the necessary decisions to deal with this complex new problem as successfully as they have done with their initial disarmament task?

The high level Advisory Panel that reported recently on the future priorities for the OPCW put the point very clearly in its early paragraphs:[16]

> 4.... the OPCW needs to prepare for a transition from mandates and efforts primarily characterised by the elimination of chemical weapons stockpiles and production facilities to an agency that will have as its main task to ensure that the menace of chemical warfare and the use of toxic chemicals for hostile purposes will never reappear, and that international cooperation and assistance in the field of peaceful uses of chemistry can flourish.
>
> 5. It is now time for States Parties and the OPCW collectively to begin addressing this transition.

The Advisory Panel also made it clear that the Convention will have to make this transition in a complex new security environment and that one of the challenges is that 'science and technology are advancing at an astounding pace, creating new opportunities but also new risks'.

Four challenges related to science and technology are very pertinent here: how to prevent Article II.9 (d) of the Convention being viewed as a licence to develop novel incapacitating chemical agents;[17] how to extend the verification system to cover the biological synthesis of chemicals (such as peptides) in large quantities;[18] how to restrain the development of systems for the long range and larger quantity delivery of riot control agents (that might then also serve to deliver other chemical agents, like novel incapacitating agents);[19] and what to do about the woeful ignorance of chemists (and scientists in other related fields, such as neuroscience) of the CWC and their responsibilities under the Convention.[20] Obviously, the focus of attention in examining the response to these challenges should be the work of the Scientific Advisory Board and, in particular, its reports to the three Five-Year Review Conferences in 2003[21], 2008 and 2013 and what impact these had on the outcomes of these reviews.

Three of these issues were clearly recognised by the SAB in its report for the First Review Conference in 2003. In regard to incapacitants the report states:[22]

> 3.14 The SAB was also aware of concerns about the development of new riot control agents (RCAs), and other so-called 'non-lethal' weapons utilising certain toxic chemicals (such as incapacitants, calmatives, vomiting agents, and the like)...The SAB noted that the science related to such agents is rapidly evolving, and the results of current programmes to develop such 'non-lethal' agents should be monitored and assessed in terms of their relevance to the Convention.

In regard to production capabilities the negotiators of the Convention had coined the term 'other chemical production facilities' (OCPFs) to cover facilities that did not process or consume any of the chemicals on verification schedules 1, 2, and 3, but which may be capable of producing such chemicals. Additionally, it was not possible for the negotiators to reach agreement on whether OCPFs included only production by chemical synthesis or also included those using biochemical processes. The term production by synthesis was used but could be understood either way.

These production issues were very difficult to resolve, but the SAB had clear views in 2003:[23]

> With the increasing globalisation of the industry, there is a need to review the verification regime for OCPF plant sites, to ensure that this

> regime is effective in monitoring the relevant parts of the chemical industry. There would appear to be a need for conducting a larger number of inspections at OCPF facilities than in previous years, because there are some OCPF facilities that are highly relevant to the Convention.

And:[24]

> 4.3 In relation to the production by synthesis of discrete organic chemicals, the SAB concluded that, from a scientific standpoint, it is no longer possible to make a clear distinction between 'chemical' and 'biological and biologically mediated' processes. The emphasis should be on the product rather than on the process.

The report continued by noting 'that view was not shared by a meeting of government experts'.

In regard to education the report also presented a very clear-cut view:[25]

> 9.2 Greater efforts in terms of education and outreach to the worldwide scientific and technical community are needed in order to increase awareness of the Convention and its benefits...
>
> 9.4 The SAB was convinced that efforts in the area of education and outreach are important to further the objectives of the Convention: these efforts include raising awareness, assuring that the principles of the Convention become firmly anchored in professional ethics and teaching, and promoting international cooperation in the field of chemistry.

The *Report* of the First Review Conference noted these observations of the SAB and requested that relevant sections of the OPCW study them further with a view to preparing recommendations on what actions might be taken.[25]

One experienced commentator concluded his report on the Second Review Conference of the CWC in 2008 by recalling the difficult atmosphere:[26]

> This Conference was more politicized than earlier meetings, which found pragmatic solutions to real problems. Perhaps this comes from realization that the destruction period is coming to an end... and that the OPCW will be inevitably changed because of this.

The conference had, in fact, been prepared for carefully with an Open-Ended Working Group (OEWG) meeting from July 2006, but its suggested text could not be developed and agreed easily. This problem eventually led to a group of 20 States Parties being invited to an Other Meeting to develop the final text of the report of the Conference with little input from States Parties not in this Other Meeting group – a manoeuvre that undoubtedly led to ill feelings. Indeed, one State Party publicly made clear its view that this procedure should not be seen as setting a precedent.

It is clear from this report of the Review Conference that no agreement on the question of OCPFs[27] was possible, even whilst 'the number of facilities handling scheduled chemicals is in the hundreds, the number of declared OCPFs is now over 4,500 worldwide'. The question of what to do about incapacitants was raised in an official paper by Switzerland,[28] but nothing appeared on the subject in the Report of the Review Conference,[29] and education and awareness-raising was almost totally ignored.[30]

Part of the input to the SAB report to each CWC Review Conference, as we have seen, is a scientific meeting organised by the International Union of Pure and Applied Chemistry (IUPAC) and a detailed report of these scientific meetings has also been published on each occasion. The meeting for the Second Review Conference was held in Zagreb[31] in 2007 and many of the issues raised there found their way into the SAB report for the Review Conference.[32] The topics of incapacitating chemicals, OCPFs and education and awareness-raising programmes were dealt with in some detail, but one interesting issue (that was covered by specific contributions in Zagreb) concerned delivery systems.

The SAB noted, for example, that:[33]

> 2.8 Many of the considerations that promote the design of particles for the effective and targeted delivery of drugs via the respiratory system would be applicable to the dissemination of a chemical-warfare agent in an aerosol. The efficiency of absorption has been improved, for instance, through the use of large porous particles that allow the delivery of drugs into the deep alveolar regions of the lungs and promote their absorption there. The spray-drying equipment needed to create such particles is relatively inexpensive and widely available.

Furthermore, the report pointed out that:

> This type of technology can be combined with nanotechnology to deliver nanoparticulate aggregates that will, once absorbed, disperse in the body, where their design (e.g. multifunctional polymeric design) could facilitate improved and more-selective delivery and targeting of drugs.

Little wonder, therefore, that the SAB saw good reason[34] to 'assess carefully the compatibility with the Convention of the development of weapons that employ toxic chemicals for law enforcement purposes (including so-called non-lethal weapons).'

The report of the SAB for the Third Review Conference stated that in its view:[35]

> the technical discussion on the potential use of toxic chemicals for law enforcement purposes has been exhaustive...The SAB recommends that the Secretariat start preparations for verification activities, relevant to incapacitating chemicals, that could be required in an investigation of alleged use (IAU).

Again, particularly at the instigation of Switzerland, this issue was discussed at the conference, but no specific language was agreed in the report of the review.[36]

In regard to OCPFs, as one commentator[37] noted in the run-up to the Review Conference, this issue 'has been a bone of contention among CWC states parties during the past decade' and at the present rate of inspection 'it will take the OPCW at least another 20 years to visit all the remaining facilities in this category just once'. In regard to production by synthesis, the SAB reaffirmed its previous recommendation[38] that this 'should include biologically mediated processes', but again there was little sign of decisive action being taken at the Review Conference.

For the first time, in its section on 'Technologies for delivery of toxic chemicals and drugs' the SAB noted:[39]

> with concern isolated reports of the commercial availability of munitions apparently designed to deliver large amounts of riot control agents over long distances.

And it reported[40] that 'in view of the important role that education and outreach play in chemical safety and chemical security' it had 'convened a TWG [Temporary Working Group] on education and outreach'. It is therefore possible to be optimistic that the educational level of chemists will be given careful consideration in the next few years. Also, the debate on incapacitants led some States to make their positions on this issue clear. Germany, for example, stated:[41]

> Germany, in her implementing legislation has deliberately opted for a narrow interpretation of the Convention. Thus, Germany has explicitly restricted toxic chemicals for law enforcement purposes that are permitted for use, both by its military and police forces, to riot control agents as defined in paragraph 7 of Article II of the Convention.

That is, to chemicals that 'can produce rapidly in humans sensory irritation or disabling physical effects which disappear within a short time following termination of exposure'. Yet, despite such progress, it seems likely that the CWC, even with greater capabilities than the BTWC, will continue to have difficulty in adjusting effectively to the rapid rate of scientific and technological change in areas such as neuroscience.

Conclusion

It is easier to make an assessment of the progress in dealing with science and technology in regard to the BTWC than the CWC, as its Seventh Review Conference took place longer ago. Many states made proposals as to how the Intersessional Process should be reshaped in the run-up and at the review, but it seems clear that no adequate mechanism to enable a sensible review of these suggestions was available for suitable text to be developed. As Pearson and Sims commented:[42]

> the proposal tabled by China, India, Iran, Pakistan and Russia on the first of the last four days on Monday 19 December 2011 ... came very late ... The subsequent proposal made on the afternoon of Tuesday 20 December 2011 by the JAKSNNK group (Japan, Australia, Canada, Republic of Korea, Switzerland, Norway, New Zealand) entitled *Draft proposal on the intersessional programme 2012–2015* just 48 hours before the inflexible termination deadline ... was even later. Late proposals with little time for consideration are not helpful for achieving a successful outcome.

No doubt there were political differences that framed the different proposals, but perhaps also a little more time at least would have produced a less unwieldy agenda for the SAI on science and technology.

We can gain an idea of what might have been possible from two Working Papers for the Review Conference, both presented by South Africa. In the first it argued for better planning in regard to the Implementation Support Unit (ISU) suggesting that:[43]

> the ISU budget and structure for the last five years was based on assumptions rather than proper planning, which resulted in an underestimation of activities as well as costs.

So it proposed that:

> (a) the Review Conference decides on the functions of the ISU and based on those functions determines a budget until the end of 2012,
>
> (b) the Review Conference instructs the ISU and interested States Parties to do detailed planning with relation to the structure and budget for 2013 till the next Review Conference for approval by the Meeting of States Parties at the end of 2012.

If that suggestion had been adopted there would have been a growing link between what States Parties decided to try to achieve in the Intersessional Process and the administrative support for those objectives. That link was not established by the Review Conference and will probably not be established until 2016 at the earliest.

In regard to the programme for the ISP, in South Africa's opinion it was necessary for work to be done on specific substantial issues and that:[44]

> this will require that the Meeting of States Parties (MSP) have decision making powers while the experts meeting should concentrate on examining specific issues for the MSP to decide upon.

The paper suggested that the Meeting of Experts should be extended from one to two weeks and that extended coverage be given in open-ended committees to three topics, one of which would be science and technology and, ideally, that these committees would be chaired by the same person for the whole of the period, thus providing the necessary continuity required for real progress to be made.

The Meeting of Experts would also discuss a couple of topics, such as revision of the CBM system or what compliance means in practical terms, as decided by the Review Conference. Reports from the Meeting of Experts would then be considered by the Meeting of States Parties[45] to 'decide on actions, continuation of work or referral to the next Review Conference for decision'. Similarly detailed proposals to increase the efficiency and effectiveness of the ISP were made by other States Parties in Working Papers for the Review Conference.

The question that remained in regard to the BTWC was whether the chairs of the 2013, 2014 and 2015 meetings could move the process towards these kinds of more productive formats. In regard to the CWC, perhaps the best indicators of progress would be the SAB Temporary Working Groups, such as that dealing with education.[46] As can be seen from the OPCW website this work on education has begun to bear fruit, with specific educational resources for teachers and students being prepared and put online for wide use by the organisation in 2014.

References

1. Secretary of State (2002) *Strengthening the Biological and Toxin Weapons Convention: Countering the Threat from Biological Weapons*. Cm 5484, HMSO, London, April.
2. Rhodes, C and Dando, M. R. (2007) Options for a scientific advisory panel for the Biological Weapons Convention, pp. 95–114 in B. Rappert and C. M. McLeish (Eds), *A Web of Prevention: Biological Weapons, Life Sciences and the Governance of Research*. London: Earthscan.
3. First Review Conference of the States Parties to the Convention on the Prohibition of the Development, Production and Stockpiling of Bacteriological (Biological) and Toxin Weapons and on Their Destruction (1980) *Final Declaration*. BWC/CONF.I/10, United Nations, Geneva, Switzerland. (p. 2).
4. Second Review Conference of the States Parties to the Convention on the Prohibition of the Development, Production and Stockpiling of Bacteriological (Biological) and Toxin Weapons and on Their Destruction (1986) *Final Declaration*. BWC/CONF.II/13/II, United Nations, Geneva, Switzerland. (p. 2).
5. Walker, R. A. (2011) *Manual for Delegates: Conference Process, Procedure and Negotiation*. United Nations Institute for Training and Research, Geneva.
6. United States (2011) *The Next Intersessional Process*. United Nations, Geneva.
7. Pearson, G. S. and Sims, N. A. (2012) *The BTWC Seventh Review Conference: A Modest Outcome*. Review Conference Paper No. 31, University of Bradford, March.
8. Seventh Review Conference of the States Parties to the Convention on the Prohibition of the Development, Production and Stockpiling of Bacteriological (Biological) and Toxin Weapons and on Their Destruction (2012) *Final Declaration*. BWC/CONF.VII/7, United Nations, Geneva, Switzerland, 13 January (pp. 23–24).
9. Seventh Review Conference of the States Parties to the Convention on the Prohibition of the Development, Production and Stockpiling of Bacteriological (Biological and Toxin Weapons and on Their Destruction (2011) *New scientific and technological developments relevant to the Convention*. BWC/CONF.VII/INF.3/Add.1. United Nations, Geneva, Switzerland, 23 November (pp. 29–31).

10. United Kingdom (2012) *The convergence of chemistry and biology: Implications of developments in neurosciences.* Working Paper No.1, BWC/MXP/WP.1. United Nations, Geneva, 12 July (p. 3).
11. South Africa (2012) *The intersessional process: Comments and proposals.* Working Paper No. 7, BWC/MSP/WP.7. United Nations, Geneva, 5 December (p. 2).
12. Pearson, G. S. and Sims, N. A. (2013) *Review No. 37: Report from Geneva: The Biological Weapons Convention Meeting of States Parties December 2012.* Harvard/Sussex Programme, University of Sussex, March (p. 38).
13. Australia *et al.* (2011) *Possible approaches to education and awareness-raising among life scientists.* Working Paper No. 20, BWC/CONF.VII/WP.20/Add.1. United Nations, Geneva, 1 November (p. 2).
14. Novossiolova, T. and Pearson, G. S. (2012) *Biosecurity Education for the Life Sciences: Nuclear Security Education Experience as a Model.* Briefing Paper No. 5, (Third Series), University of Bradford, October.
15. Ambassador J. Körömi (2013) *Biological Weapons Convention Meetings in 2013.* Secretariat, BWC Implementation Support Unit, United Nations Office for Disarmament Affairs (Geneva Branch), Palais des Nations, CH-1211, Geneva 10, Switzerland, 18 February.
16. Director General (2011) *Report of the Advisory Panel on Future Priorities of the Organization for the Prohibition of Chemical Weapons.* S/951/2011, OPCW, The Hague (pp. 3–4).
17. International Committee of the Red Cross (2012) *'Incapacitating Chemical Agents': Law Enforcement, Human Rights Law and Policy Perspectives.* Experts Meeting, Montreux, Switzerland, 24–26 April. ICRC, Geneva.
18. Matthews, R. J. (2009) The regime for other chemical production facilities: A technical perspective. *The Chemical and Biological Weapons Conventions Bulletin,* **83** + **84**, 5–13. Harvard/Sussex Programme, University of Sussex, July.
19. Crowley, M. (2013) *Drawing the line: Regulation of 'wide area' riot control agent delivery mechanisms under the Chemical Weapons Convention.* Omega Research Foundation and University of Bradford, April.
20. Scientific Advisory Board (2013) *Report of the Second Meeting of the SAB Temporary Working Group on Education and Outreach in Science and Technology Relevant to the Chemical Weapons Convention.* SAB-20/WP.1, OPCW, The Hague, 25 February.
21. Conference of the States Parties (2003) *Report of the First Special Session of the Conference of the States Parties to Review the Operation of the Chemical Weapons Convention (First Review Conference) 28 April–9 May 2003.* RC-1/5, OPCW, The Hague, 9 May (p. 9).
22. ibid, p. 15.
23. ibid, p. 17.
24. ibid, p. 16.
25. ibid, p. 26.
26. Guthrie, R. (2008) The Second Chemical Weapons Review Conference. *The Chemical and Biological Weapons Conventions Bulletin,* **79**, 1–6, The Harvard/Sussex Programme, University of Sussex, June, (p. 4).
27. ibid, p. 2.
28. Switzerland (2008) *Riot Control and Incapacitating Agents under the Chemical Weapons Convention.* RC-2/NAT.12, OPCW, The Hague, 9 April.

29. Conference of the States Parties (2008) *Report of the Second Special Session of the Conference of the States Parties to Review the Operation of the Chemical Weapons Convention (Second Review Conference) 7–18 April*. RC-2/4, OPCW, The Hague.
30. ibid, p. 29.
31. Balali-Mood, M. *et al.* (2008) Impact of scientific developments on the Chemical Weapons Convention. *Pure Appl. Chem.*, **80** (1), 175–200.
32. Director-General (2008) *Report of the Scientific Advisory Board on Developments in Science and Technology*. RC-2/DG.1, OPCW, The Hague, 28 February.
33. ibid, p. 11.
34. ibid, p. 15.
35. Conference of the States Parties (2012) *Report of the Scientific Advisory Board on Developments in Science and Technology for the Third Special Session of the Conference of the States Parties to Review the Operation of the Chemical Weapons Convention*. RC-3/DG.1, OPCW, The Hague, 29 October (p. 4).
36. Conference of the States Parties (2013) *Report of the Third Special Session of the Conference of the States Parties to Review the Operation of the Chemical Weapons Convention*. RC-3/3, OPCW, The Hague, 19 April.
37. Kelle, A. (2013) The Third Review Conference of the Chemical Weapons Convention and beyond: Key themes and prospects for incremental change. *International Affairs*, **89** (1), 143–158 (p. 147).
38. Reference 35, p. 11.
39. Reference 35, p. 14.
40. Reference 35, p. 31.
41. Germany (2013) *Toxic Chemicals for Law Enforcement*. RC-3/NAT.44, OPCW, The Hague, 16 April (p. 2).
42. Reference 7, pp. 135–136.
43. South Africa (2011) *Biological Weapons Convention Support Unit: Future Planning*. BWC/CONF.VII/WP.17, United Nations, Geneva (p. 2).
44. South Africa (2011) *Proposal for the Intersessional Process*. BWC/CONF.VII/WP.18, United Nations, Geneva.
45. ibid, p. 2.
46. Scientific Advisory Board (2014) *Report of the Third Meeting of the Scientific Advisory Board's Temporary Working Group on Education and Outreach in Science and Technology Relevant to the Chemical Weapons Convention*, SAB-21/WP.3. OPCW, The Hague, 7 January.

11
Where Are We Going?

Introduction

Before considering what might best be done to minimise the possible misuse of advances in neuroscience for hostile purposes, it is useful to take stock briefly of where we are today and where we are likely to be going.

The CWC

In June 2014 the Organization for the Prohibition of Chemical Weapons published the report of the Temporary Working Group (TWG) of the Scientific Advisory Board on the convergence of chemistry and biology.[1] The TWG set out its objectives, in part, as follows:[2]

> Many articles published on the implications for the CWC of rapid advances in the life sciences, including the convergence of chemistry and biology, have approached the topic from the perspective of what is theoretically possible. Few articles have discussed what is realistically possible to achieve with the current state of the development of science and technology.

The report also acknowledged that there had been interest in the possible implications of advances in the life sciences for the BTWC and that there was a joint interest in examining this issue in regard to mid-spectrum agents, toxins and bioregulators. But the main thrust of the argument on objectives was clearly that a realistic analysis of current practical possibilities was now required.

In this regard the report came to some, apparently very clear, conclusions, for example, with respect to toxins it suggested that:[3]

> Although the technical capability to chemically synthesize many toxins exists today, there are practical limitations with regard to scale and complexity. *The threat of possible misuse of this technology with regard to the CWC is therefore currently considered low.* [emphasis added]

And in regard to bioregulators the report stated:[4]

> The potential of peptides for development as incapacitants may have been overstated by some commentators. Peptides could be produced using metabolic engineering and synthetic biology but the pharmaceutical industry currently regards chemical synthesis, using specialized equipment, as the most cost-effective method for producing many small peptides. *The threat of possible misuse of this technology with regard to the CWC is currently considered low.* [emphasis added]

Even in regard to nanotechnology the report stated:[5]

> Nanotechnology is playing an important role in improving drug delivery to the body, protective equipment, and in the development of biosensors. Nanotechnology *may* have *some* potential for application to purposes prohibited by the CWC. [emphases added]

Overall then, on first reading the tenor of the report is reassuring. We really do not appear to have much that we need to worry about, at the moment, in regard to mid-spectrum agents and the convergence of chemistry and biology.

Coincidentally, in 2014, the open access journal *Toxins*[6] carried a long, detailed article titled *Overall View of Chemical and Biochemical Weapons.*[7] In concluding the author argued that the development of chemical weapons using traditional technologies had reached its limits, but that there was a big potential for further development, particularly because of the convergence of chemistry and biology. He suggested that there is a high risk because[8] 'today, there is a general acceptance of the development of non-lethal chemical weapons ... at a technologically higher level' and outlined some of the relevant advances in this area. The author, who works in the biomedical engineering department of a major European university, was not sanguine about the future prospects of effective verification of compliance with the BTWC and CWC control regimes.

That kind of specialist viewpoint being published in 2014 implies that a more careful second look at the details of the TWG report on convergence might well be warranted. It is important to note that the TWG's remarks in regard to toxins and bioregulators have the significant word *'currently'* in both conclusions and that, in regard to nanotechnology, there is a clear acknowledgement that some advances may be of concern in relation to the CWC. It is hardly surprising that the TWG's recommendations point to the need for continuing monitoring and assessment of these aspects of convergence (Table 11.1). What then might be of concern, and what indicators are there in the TWG's full report?

The group was asked to examine the possible use of biologically mediated processes for the production of toxins (rather than chemical synthesis). They noted, for example, that:[9]

> The natural metabolic pathways have been published for the production of saxitoxin, ricin and many other toxins. *In vitro* biosynthesis of ricin and saxitoxin has been described.

and:

> Awareness and concerns about ricin have been growing as a result of incidents of misuse. Recent research has focused mainly on toxicity, detection, countermeasures, and medical applications (e.g. immunotoxins for cancer therapy).

It is interesting to note that there were two invited presentations on toxins at the 2014 BTWC Meeting of Experts and that one of these described EQuATox, an EU funded network focused on quality assurance for the

Table 11.1 Recommendations of the TWG in relation to toxins, bioregulators and nanotechnology

The SAB, or a suitable TWG, should review the feasibility of using metabolic engineering or synthetic biology to obtain toxins prior to the next review conference.
The TS [OPCW Technical Secretariat] should increase and maintain in-house knowledge of bioregulators, and possible applications of new developments in drug delivery.
The SAB should monitor advances in nanotechnology prior to the next review conference.

Source: Modified from Organization for the Prohibition of Chemical Weapons (2014) *Convergence of Chemistry and Biology: Report of the Scientific Advisory Board's Temporary Working Group*. OPCW, The Hague, June.

detection of high priority biological toxins.[10] The toxins of concern were ricin, saxitoxin, botulinum neurotoxins and Staphylococcal Enterotoxin B. The EQuATox's funding, which was for three years from the start of 2012, totalled EU 1,338,634 and involved 36 laboratories from 24 countries.

With respect to peptides, the TWG report argues that whilst some peptides like Substance P could be a worry, because of the induction of bronchoconstriction, no centrally acting peptide has been the subject of published data that suggests it could be of concern. However, as the report notes:[11]

> The shortcomings of peptides as drugs (and by implication, for uses prohibited by the Convention) can be moderated in several ways. Formulations, particularly associated with liposomes or nanocarriers, are being explored to enhance penetration of the blood-brain barrier, overcome host defences, and target specific organs.

The report continues:

> Drug companies have screened large numbers of metabolically resistant analogues, mostly with unnatural (chemical) modifications, some of which may have substantially increased potency and toxicity.

The report suggests that such modifications may, however, increase the complexity and costs of the end product. Nevertheless, nanotechnology could be of considerable use to the weaponeer:[12]

> Nanocarrier-based delivery systems present several advantages over the classic ones: overcoming solubility problems, protecting the drug from the external environment (temperature, UV radiations, pH), and controlling the release profile.

Furthermore:

> Nanocarrier-based delivery systems permit a more precise and controlled targeting at the site of action, while reducing the time of exposure of non-targeted tissues.

This perception of the kinds of advantages that nanotechnology is likely to bring – for good or ill – is widely shared in other detailed analyses.[13]

In short, whilst there may be differences of opinion about the impact of advances in relevant science and technology on the threat of use of chemical and biological agents today, there is a widespread understanding that these advances could be misused in coming decades. Thus, the optimistic scenario is that we still have time to put in place sensible policies that will minimise these threats whilst protecting benignly intended science and technology in the future. So, how well is the development of the BTWC progressing? In particular, was much more significant progress made in the intersessional process meetings in 2013 and 2014 than had been possible in 2012, immediately after the 2011 Seventh Review Conference?

The 2013 BTWC meetings

A close look at the States Parties' papers for the Seventh Review Conference shows clearly that there was a wide consensus amongst major states for making a much more efficient and effective review of science and technology.[14] India, for example, proposed that:[15]

> the Seventh Review Conference take a decision regarding structured and systematic review of S & T developments within the framework of the Convention. The aim is to build consensus among Member States based on a thorough review of developments in life sciences and biotechnology that are relevant to the BWC.

and China stated:[16]

> In order to promote greater flexibility and efficiency, China supports making appropriate improvements in the intersessional process currently in use. Experts meetings can continue to be held, or working groups open to all States Parties can be set up, to carry out specialized discussions of such topics of broad concern as ... assessment of the impact of scientific and technical development ... Results of the discussions should be made available to the meetings of States Parties, which would in turn submit a report on them to the subsequent Review Conference.

A number of other major states made suggestions along similar lines so there was a possibility of agreement on a significant improvement in the review of relevant science and technology.

Some of these papers went into detail on how a better review system for science and technology might be structured. For example, a paper by Australia, Japan and New Zealand suggested a systematic five-step process.[17] In the first step, the Review Conference or the Meeting of States Parties in the Intersessional Process (ISP) would decide on one or more science and technology topic(s) to be the subject of review in the following year. Then the Meeting of Experts in the following year would invite various expert groups, such as the InterAcademy Panel (IAP), to prepare factual reviews of the topic(s). These experts would discuss their factual reports with representatives of States Parties at the Meeting of Experts (which is held in the middle of the year). The States Parties would then consider the implications in sessions and a facilitator, appointed for the whole of the ISP between the five-year review conferences, would prepare a report reflecting the factual reviews, and the views of States Parties, of the implications for the BTWC. These views would not necessarily be a consensus. The facilitator's report would be circulated to States Parties well in advance of their meeting at the end of the year and that meeting would decide on any *actions* that might need to be taken. Such actions would be subject to review at the next five-year conference. Finally, the cycle would recommence with the Meeting of States Parties deciding on the topic, or topics, that would be reviewed in the following year.

Certain aspects of this proposal stand out. The number of topics would be limited and there would be time for considered scientific input to be prepared for the Meeting of Experts. This meeting would have time to discuss the factual reviews with experts in the field and to consider the implications for the Convention. Crucially, a facilitator with a remit covering the whole of the five-year ISP period would prepare a report on the factual reviews and on the implications, including differing views. This report would be made available in good time for States Parties to consider before the Meeting of States Parties which would then *decide* on what *actions* needed to be taken in regard to the chosen topic. Cumulative experience of the process, and of the actions, would then be analysed at the next review conference, allowing systematic improvements to be made if they were considered necessary.

In the event, a sensible system like this could not be agreed at the Seventh Review Conference and, as we have seen, the system that was agreed had too many topics and too little time allocated to make it simple to achieve cumulative progress. Nevertheless, some States Parties pressed on with efforts to make such progress. One paper, by seven States Parties, argued that to promote effective action it would be necessary for the 2013 Meeting of Experts to identify and address issues where:[18]

> There is something new to say... There is enough agreement that something specific can be said [and] there is something for States Parties to do.

In short, it was, in this view, simply not possible to just repeat bland, agreed statements that had appeared in the reports of previous ISP meetings.

Regrettably, what happened, despite the strenuous efforts of many delegates and officials in 2013, was that numerous good proposals were lost in the process of crafting the final report of the Meeting of States Parties. For example, the Chair of the meeting had proposed a series of quite specific actions that might be included in the report, such as considering:[19]

> (c) A board to provide scientific advice, similar to the Scientific Advisory Board of the CWC, or based on a different model;
>
> (d) An open-ended working group to consider the implications of advances in technology, including the convergence of chemistry and biology;
>
> (e) Encouraging States Parties that host national or international meetings addressing relevant science and technology developments to prepare a summary on the implications for the Convention.

None of these suggestions reached the final report. It might, of course, be argued that none of these suggestions *required* States Parties to carry out actions, but they were surely pointing the way towards actions that could lead to a better, more effective, review of science and technology relevant to the Convention.

It is difficult to pin down exactly why States Parties have failed to give more serious attention to advances in science and technology in the ISP meetings since the Seventh Review Conference in 2011 or, for that matter, to improve the overall efficiency and effectiveness of their meetings. A variety of possible explanations has been put forward by commentators.[20] These range from differences between states about the possibility of verification of compliance, and therefore a reluctance to seriously discuss the implications of advances in science and technology, through to the BTWC still not being seen as a major priority in many capitals, and therefore the likelihood that outside events will disrupt BTWC meetings and preparations for meetings. One thing that is pretty obvious to outside observers is that there is little expectation of rapid progress in strengthening the BTWC and therefore a tendency to set ever-diminishing objectives for the BTWC meetings. The problem is compounded because very few diplomats have a background in science and technology and, despite valiant efforts to bring scientists to the meetings in order to make expert presentations, few scientists have the experience and capability to communicate complex scientific issues to non-scientists. Thus, as South Africa noted after the 2012 Experts Meeting, there is little detailed productive discussion of science and technology in the sessions on science and technology.[21]

That is not to say that all States Parties were content to continue without making more effort to accelerate progress in the run-up to the 2016 Eighth Review Conference of the BTWC. At the December 2013 Meeting of States Parties in Geneva a paper by 12 States from diverse parts of the world argued that:[22]

> BWC States Parties must continue to engage in constructive discussions with a goal of promoting *effective actions* on the implementation and enforcement of all aspects of the BWC. [emphasis added]

This point about the need for *effective action* was made by the Chairman of the 2014 meetings in a letter to States Parties in the following February. In his view, one way to move forward was:[23]

> to start giving greater focus to the *effective action* part of our mandate. The reports of the 2012 and 2013 Meetings of States Parties contain a broad range of *common understandings* some quite detailed. In 2014 we will continue to discuss and promote common understandings. [emphases added]

But, he added:

> as we move closer to the Eight Review Conference in 2016, this may be a suitable point to turn more of our attention towards options for promoting effective action.

The need for effective action could not have been clearer given that the States Parties had been unable to do anything about the ongoing gain-of-function experiments with deadly viruses.

This air of unreality and detachment from the issue, in a place that should have been the centre of a discussion on the dual-use problem, continued in the BTWC Experts Meeting of August 2014. There, the United States presented a paper titled, *The United States of America policy for oversight of life sciences dual use research of concern (DURC)*.[24] The paper noted that a 2012 policy 'requiring U.S. federal departments and agencies that fund life sciences research to identify and manage the risks associated with dual use research of concern (DURC)' would shortly be complemented by a 'second policy that expands DURC oversight to research institutions receiving U.S. federal funding.' The paper then suggested that:[25]

> Together, the two U.S.A. Government DURC oversight policies work to engage life science research institutions and federal funding agencies in a shared responsibility to address the risk that the knowledge, information, products, or technologies generated from the life sciences could be used for harm.

Now it has to be said that the United States has done more than most countries to deal with the problem of dual use and biosecurity. However, the paper makes no mention of the view of the Fink Committee in 2004 (see Chapter 4) that such actions in the United States would need to be complemented by similar action by other states and international bodies to adequately deal with the dual-use problem.

Moreover, before the BTWC meeting in Geneva in August 2014 a series of major errors in biosafety in key institutions in the United States had clearly raised the question of whether *any* institution could be relied on to keep safe the kind of deadly contagious influenza strains being developed in gain-of-function experiments.[26] A letter from senior congressional representatives to the Director of the US National Institutes of Health[27] raised a series of questions about these biosafety lapses, for example, the finding of decades-old, but still viable, vials of smallpox at a Food and Drug Administration laboratory! However, the letter ended with a question about the NSABB, which had been set up after the report of the Fink Committee early in the new millennium. The question was preceded by a paragraph which stressed the continued importance of the National Science Advisory Board for Biosecurity in the light of recent lapses in federal laboratories and noted that the Board had not met for two years and that its charter had been changed in March 2014 'so that the Board no longer has responsibility to review research that might be repurposed for bioterrorism or bioweapons'. It then continued, alarmingly, to state that:[28]

> 11 of the 23 panel members were suddenly dismissed by NIH official and executive director of the NSABB, Mary Groesch, on a Sunday evening (13 July 2014).

The letter stated that this raised serious questions for the House Committee on Energy and Commerce and that, to assist their enquiries, they would like to have by mid-August 2014:

> All emails since January 1, 2012 in the possession of Mary Groesch relating to the NSABB, including the dismissal of the 11 members of the Board, the change in the Board's charter, and why the Board has not met in nearly two years.

Early in 2012, of course, was the start of the acrimonious debate on gain-of-function experiments funded by the National Institutes of Health. So what was happening in civil neuroscience as the slow-motion activity continued in Geneva?

Modern civil neuroscience

There had been concerns expressed that the EU Human Brain Project (HBP) was trying to produce a simulation of the brain without adequate data on brain functions being available to carry out that task. These concerns surfaced in an open *Message to the European Commission concerning the Human Brain Project*[29] in July 2014. According to the authors of the message:

> the HBP has been controversial and divisive within the European neuroscience community from the beginning. Many laboratories refused to join the project when it was first submitted because of its focus on an overly narrow approach, leading to a significant risk that it would fail to meet its goals. Further attrition of members during the ramp-up phase added to this narrowing.

The message went on to note that there was an ongoing review taking place and suggested that some tough criteria – for example, on transparency and independence – would be necessary for the review to be successful.

If the review was not satisfactory, the message called for a reallocation of funding and if the criteria were not accepted by the EU, the 759 signatories and supporters (on 28/08/2014) pledged not to apply for HBP partnering projects and to urge their colleagues to do the same.[30] All was clearly not well with the EU Human Brain Project in mid-2014. However, the leaders of the project have made it clear that the critics have simply misunderstood the nature of the project. In their view, there is more than enough data being produced and that what is needed is a paradigm shift – to move to integration of all the data – and hence the need for the priority given to information technology developments.[31]

The final report of the Working Group to the Advisory Committee to the Director of the US National Institutes of Health was published on 5 June 2014.[32] This suggested a ten-year programme for research and development be undertaken:[33]

> we recommend an investment by the NIH that ramps up to $400 million/year over the next five years (FY16–20), and continues at $500 million/year subsequently (FY21–25).

If implemented on this scale, the US programme would clearly significantly dwarf the EU programme in scale.

The report again emphasised the importance of understanding and manipulating brain circuits (see Chapter 5):[34]

> the working group identified the analysis of circuits of interacting neurons as being particularly rich in opportunity with potential for revolutionary advances.

And the group also appeared to embrace the need for a kind of paradigm shift, with the ten-year programme divided into five years focused on technology development and five years focused on 'integrating technologies to make fundamental new discoveries about the brain'.

The objective of this work was quite clearly set out:[35]

> The most important outcome of the initiative will be a *comprehensive mechanistic understanding* of mental function that emerges from synergistic application of the new technologies and conceptual structures developed under the BRAIN initiative. [emphasis added]

There remained ethical sections of the report, but again, as in the interim report, there was no mention of the possibility of dual use or the hostile misapplication of such benignly intended work.

What was noted, however, was the possibility of radical new findings:[36]

> we also emphasise the likelihood of entirely new, unexpected discoveries that will result from the new technologies. In some sense, BRAIN initiative scientists who apply the new activity-monitoring technology will be like Galileo looking at the heavens with the first optical telescope.

The questions then are, what might they find and how can it be kept free from hostile applications?

We can get an idea of the kind of information likely to become available from recent work on the manipulation of memory in mice. Building on previous detailed work,[37] the authors of a paper in *Nature* in late August 2014 were able to manipulate the memory circuits in mice to reverse a bad memory. As they stated:[38]

> We found that in the DG [dorsal dentate gyrus of the hippocampus], the neurons carrying the memory engram of a given neutral context have plasticity such that the valence of a conditioned response evoked by their reactivation can be reversed.

It is well known that memories are not static but can be modified when they are reactivated but, as the authors point out, their 'present work provides new insights into the functional neural circuits underlying the malleability of emotional memory'.

Military interest

In the UK Royal Society's *Brain Waves Module 1: Neuroscience, society and policy* Professor Andy Stirling contributed an essay on 'Governance of neuroscience: challenges and responses'. In this essay he suggested that over recent decades debates about the governance of new technologies have provided some hard-won lessons. Amongst these lessons was:[39]

> *See no evil*
>
> A particular technology may realise its initial promise, but this very feasibility may itself create opportunities for deliberate or inadvertent misuse.

Moreover:

> Although readily foreseeable in the same terms as benign uses, malign applications are typically understated in regulatory assessments... Yet

> easily anticipated effects may be of a magnitude that seriously jeopardises overall benefits.

This, Stirling maintains, is well exemplified by the fact that military aims are so under-scrutinised when some one-third of the world's research and innovation is directly or indirectly designed 'to refining ways to perpetrate premeditated organised violence'.

Of course, that is not to suggest that we should stop trying to help people suffering from mental illness, but Stirling argues that with an uncertain proportion of world neuroscience research devoted to military ends, and the clear dual-use implications of work in the chemical and biological sector, the potential military implications of even ostensibly non-military applications of neuroscience should be kept in mind. In short, precaution should be the watchword.

The DARPA website carried a news item in mid-2014 titled 'Journey of discovery starts towards understanding and treating networks of the brain'. This described work that was to begin on DARPA's Systems-based Neurotechnology for Emerging Therapies (SUBNETS) programme at UC San Francisco and Massachusett General Hospital.[40] A UC San Francisco press release explained that the project was to cost some $26 million as part of the US BRAIN Project and DARPA was involved because of the disproportionate numbers of soldiers suffering psychiatric conditions.[41] A longer article explained that the objective of the work was:[42]

> first to identify brain signalling pathways specifically associated with anxiety and depression, then to develop devices to provide precise stimulation therapies that guide the brain to strengthen alternative circuits.

The research, the article explained, would build on the expertise developed in previous years on deep brain stimulation for movement disorders like Parkinson's Disease. This, it pointed out, is not just a movement disorder, but is in fact a 'neuropsychiatric disorder that includes problems with mood, thinking, anxiety, impulsivity... symptoms [that] are as fundamental a part of the disorder as slow movement or tremor'. And as these symptoms wax and wane, recordings of what is happening in the brain should reveal the cause of the waxing and waning, and how it might be corrected by appropriate, externally generated stimulation.

Another press release from UC Berkeley noted that:[43]

> Engineers from UC Berkeley, Lawrence Livermore National Laboratory and Cortera have already made headway in developing a state-of-the-art neuromodulation medical device for the project. Called OMNI the device consists of low-power, miniaturized electronics that sense and stimulate neural networks to counteract dysfunctional circuits.

The press release states that Lawrence Livermore will receive separate funding from DARPA for the project, and one of the scientists is quoted as saying, 'we obviously have many *social problems* that stem from mental illness, and I'm excited to be developing state-of-the-art electronics that contribute towards a solution.' [emphasis added]

The UC San Francisco article also states that 'all work on the new project will be conducted with the oversight of UCSF- and DARPA-based ethical committees'.[44] Given Stirling's concerns about the lessons we should have learnt already about the potential misuse of emerging technologies, and the uncertain history of dealing with dual-use issues in the life sciences, careful monitoring and caution would not seem out of place. In addition to hopes for long-term help for people clearly suffering from medical disorders we surely need to be careful to try to minimise the possibility that this work will end up being misused.

Conclusion

It is hard to avoid the conclusion that, despite the continued reporting of relevant research in the popular media[45] and the recognition of the biosecurity problem by eminent neuroscientists,[46] collectively we are a long way from grasping the opportunity to put in place effective policies to protect benignly intended work in this area from hostile misuse.

References

1. Organization for the Prohibition of Chemical Weapons (2014) *Convergence of Chemistry and Biology: Report of the Scientific Advisory Board's Temporary Working Group*. OPCW, The Hague, June.
2. ibid, p. 6.
3. ibid, p. 24.
4. ibid, p. 25.
5. ibid, p. 26.
6. See the open access journal *Toxins* <www.mdpi.com/journal/toxins>.
7. Pitschmann, V. (2014) Overall view of chemical and biochemical weapons. *Toxins*, 6, 1761–1784; doi: 10.3390/toxins 6061761.

8. ibid, p. 1781.
9. Reference 1, p. 14.
10. Meeting of Experts (4–8 August 2014) Presentation by Germany, afternoon of Monday 4 August.
11. Reference 1, p. 15.
12. Reference 1, p. 22.
13. United Nations Interregional Crime and Justice Research Institute (2012) *Security Implications of Synthetic Biology and Nanobiotechnology: A Risk and Response Assessment of Advances in Biotechnology*. Turin: UNICRI .
14. Dando, M. R. (2014) *To What Extent Was the Review of Science and Technology Made More Effective and Efficient at the 2013 Meeting of BTWC States?* Policy Paper No. 5, Biochemical Security 2030, University of Bath, May.
15. India (2011) *Proposal for structured and systematic review of science and technology development under the convention*. BWC/CONF.VII/WP.3, United Nations, Geneva, 11 October.
16. China (2011) *China's view on strengthening the effectiveness of the BWC*. BWC/CONF.VII/WP.24, United Nations, Geneva, 5 December.
17. Australia, Japan and New Zealand (2011) *Proposal for the annual review of advances in science and technology relevant to the Biological Weapons Convention*. BWC/CONF.VII/WP.13, United Nations, Geneva, 19 October.
18. Australia, Canada, France, Germany, Netherlands, the United Kingdom and the United States (2013) *Getting past yes: Moving from consensus text to effective action*. BWC/MSP/WP.4, United Nations, Geneva, 6 December.
19. Reference 14, p. 16.
20. Reference 14, p. 20, Table 4.
21. South Africa (2012) *The intersessional process: Comments and proposals*. BWC/MSP/WP.7, United Nations, Geneva, 5 December.
22. Australia *et al.* (2013) *Addressing modern threats in the Biological Weapons Convention: A food for thought paper*. BWC/MSP/2013/WP.10, United Nations, Geneva, 10 December.
23. Ambassador Urs Schmid (2014) *Biological Weapons Convention: Meetings 2014*. Letter dated 14 February. BWC Implementation Support Unit, United Nations, Geneva, United Nations.
24. United States (2014) *The United States of America Government policy for oversight of life sciences dual use research of concern (DURC)*. BWC/MXP/WP.7.Corr.1. United Nations, Geneva, 31 July.
25. ibid, p. 2.
26. Novossiolova, T. A. and Dando, M. R. (2014) Making viruses in the lab deadlier and more able to spread: An accident waiting to happen. *Bulletin of the Atomic Scientists*, http://thebulletin.org/making-viruses-lab-deadlier-and-more-able-spread-accident-waiting-happen7374. 12 August 2014.
27. Upton. F., Murphy, T., Barton, J., and Blackburn, M. (2014) *Letter to the Honorable Francis Collins, M.D.* Committee on Energy and Commerce, House of Representatives, U.S. Congress, 28 July.
28. ibid, p. 3.
29. Sample, I. (2013) Arguments over brain simulation come to a head: More than 100 scientists threaten boycott; critics say EU 1.2 bn study is premature and set to fail. *The Guardian*, 7 July.

30. Abeles, M. *et al.* (2013) *Open message to the European Commission concerning the Human Brain Project*, 7 July. Available at <http://www.neurofuture.eu/>. 28 August 2014.
31. Frackowiak, R. (2014) Defending the grand vision of the Human Brain Project. *New Scientist*, 2978, 16 July.
32. Advisory Committee to the NIH Director (2014) *Brain 2025: A Scientific Vision*. National Institutes of Health, 5 June.
33. ibid, p. 8.
34. ibid, p. 5.
35. ibid, p. 7.
36. ibid, p. 7.
37. Ramirez, S., Tonegawa, S., and Liu, X. (2013) Identification and optogenetic manipulation of memory engrams in the hippocampus. *Frontiers in Behavioral Neuroscience*, **7**, 226–239.
38. Redando, R. L. *et al* (2014) Bidirectional switch of the valence associated with a hippocampal contextual memory engram. *Nature*, **513** (7518), 426–430.
39. Stirling, A. (2011) Governance of neuroscience: Challenges and responses, pp. 87–97 in Royal Society, *Brain Waves Module 1: Neuroscience, society and policy*. Royal Society, London, January.
40. DARPA (2014) *Journey of Discovery Starts towards Understanding and Treating Networks of the Brain*. DARPA, 27 May. Available at <http://www.darpa.mil/News Events/Releases/2014/05/27a.aspx>. 27 August 2014.
41. UC San Francisco (2014) *Untangling Brain Circuits in Mental Illness: Depression, Anxiety Disorders, Addiction*. Available at <http://www.ucsf.edu/news/2014/05/114631/untangling-brain-circuits-mental-illness>. 27 August 2014.
42. Farley, P. (2014) *New Venture Aims to Understand and Heal Disrupted Brain Circuitry to Treat Mental Illness: In Support of the President's Brain Initiative Project Seeks Permanent Cures for Anxiety Disorders, Depression, Addiction*. Available at <http://www.ucsf.edu/news/2014/05/114621/new-venture-aims.illness>. 27 August 2014.
43. Yang, S. (2014) *CNEP researchers target brain circuitry to treat intractable mental disorders*. Available at <http://newscenter.berkeley.edu/2014/05/27/cnep-targets-brain-circuitry-to-treat-mental-disorders/>. 27 August 2014.
44. Reference 42, p. 3.
45. Zimmer, C. (2014) Real Zombies: The strange science of the living dead. *National Geographic*, November, 36–54.
46. Tracy, I. and Flower, R. (2014) The warrior in the machine: neuroscience goes to war. *Nature Reviews Neuroscience*, **15** (12), 825–834.

12
The Governance of Dual-Use Neuroscience

Introduction

Given the history of major misuse, for hostile purposes, of advances in science and technology, perhaps the most likely outcome over coming decades is that precisely the same thing will happen to the ongoing advances in neuroscience. Then, as Meselson[1] warned, 'therein could lie unprecedented opportunities for violence coercion, repression or subjugation'. It is important, however, to grasp that this need not necessarily happen.

In his study, *In Defence of History*, Richard Evans pointed out that:[2]

> History, then, can produce generalisations ... It can identify or point with a high degree of plausibility, patterns, trends and structures in the human past ... But history cannot create laws with predictive power.

He continued by stating that 'all those who thought, or claimed, that they had discovered laws in history were ... wrong'. And to illustrate his point, he argued that the Bolsheviks, who thought they had discovered that revolutions inevitably led to military dictatorships, proceeded to gang up on Trotsky in an attempt, in Evan's words, to 'do their level best' to break that law.

So, there is no inevitability about the future major misuse of neuroscience for hostile purposes. Whilst that may be the most likely outcome, it is not the only possibility, particularly if strenuous efforts are made by those who can see the dangers to prevent it happening. The question then is what can best be done and who can do what is needed?

A reality check

It is important, nevertheless, to be realistic about the predicament that we are in today. First, there are reasons to believe that the state system that has prevailed since the Peace of Westphalia in 1648 is under threat.[3] Some argue that the present state-based system evolved because the increased size and complexity of human societies required a new form of bureaucratic, hierarchical integration that nation states provide. That increase in the size and complexity of our societies, it is argued, came about because of the impact of advances in science and technology, which first began in Europe where the Peace of Westphalia was agreed.

Following this argument, the increase in interconnectedness, size and complexity of human societies has now, because of the continued impact of advances in science and technology, begun to force a change back to something more like the medieval – less state-centric – system. As one expert was quoted as saying:[4]

> Neo-medievalism, on the other hand, means overlapping authorities, divided sovereignty, multiple identities and governing institutions, and fuzzy borders.

From a hopeful perspective, this process could lead to a further integration of international society as multiple networks at different government and non-government levels grapple with our complex shared problems of sustainable growth and justice.

However, given the situation in late 2014, an opposing pessimistic perspective is hard to dismiss. Hopes for an Arab Spring evolving into a Middle East of stable, successful democracies looks a very distant prospect.[5] Much more likely, it seems, is that we face a generations-long period of conflict there, with its impact spreading to many other parts of the world because of terrorist actions. Moreover, the situation in Ukraine suggests that those who worried that the post-Cold War settlement in Europe was too one-sided, and might not remain acceptable to Russia for long, may well have been correct. Worrying moves towards rearmament on both sides could presage a long period of difficult relations between Russia and the United States and its allies. And, of course, that is to leave aside the difficulties that the West will have in accommodating a much more powerful China in the fast-growing Asian region.

So, the evolution of governance of science and technology – and the results of our new investigations of the brain – will almost certainly take place in a period of societal turbulence, one that will involve conflict and violence on a variety of scales. The temptation to put advances in neuroscience to use in these conflicts will always be present, and pressing reasons for immediate action can be expected from advocates, whatever the potential longer-term dangers that may arise. It is against that background that the present state and future prospects for the governance of neuroscience and its impacts have to be examined.

It is important to understand that the dangers of misuse of the life sciences are well understood, at least by the military. The UK Ministry of Defence Strategic Trends Programme's publication, *Global Strategic Trends – Out to 2040*[6] did its best to make this clear, stating:[7]

> The era out to 2040 *will* be a time of transition; this is *likely* to be characterised by instability, both in the relations between states, and in the relations between groups within states... the struggle to establish an effective system of global governance, capable of responding to these challenges, *will* be a central theme of the era.

It continued by arguing that during this period the distribution of power will change:[8]

> Out to 2040, the locus of global power *will* move away from the United States (US) and Europe to Asia, as the global system shifts from a uni-polar towards a multi-polar distribution of power.

Clearly, within such a transitional period, conflict and warfare are strong possibilities. As the document notes,[9] 'out to 2040, there are few convincing reasons to suggest that the world will become more peaceful'. And it should be noted that 'likely' is intended to convey a probability of 60 per cent to 90 per cent and 'will' a probability of greater than 90 per cent.

In regard to our central concern here, the document states bleakly:[10]

> The CBRN threat from state and non-state actors is *likely* to increase, facilitated by lowering of some entry barriers, dual purpose industrial facilities and the proliferation of technical knowledge and expertise. Terrorist attacks using chemical, biological and radiological weapons are *likely*.

Moreover, seemingly innocent advances in neuroscience cannot be assumed to be immune from misuse in such circumstances.[11]

Non-security implications of neuroscience

Robert Blank, who had written an early account of the potential implications of advances in neuroscience in the 1990s,[12] summarised the present situation in a new book in 2013 titled *Intervention in the Brain: Politics, Policy and Ethics*.[13] Following introductory chapters on the human brain, current capabilities to intervene in the operations of the brain and neuroethics and neuropolicy, Blank has a series of chapters that deal in some detail with non-security implications, such as addiction, individual responsibility, and the legal, media and commercial applications of new technologies that can be used to affect brain processes. He also deals with issues more directly related to international security, such as aggression, racism and conflict within national settings.

Blank might best be characterised as a cautious optimist. He is aware that many people could benefit from the work of neuroscientists today, but is also aware of the mistakes that have been made in the past, for example in the use of frontal lobotomy on many patients when there was considerable evidence of the dangerous nature and severe consequences of the use of this technique. As he noted,[14] this was done in spite of 'early evidence of postoperative infections and autopsies showing large areas of the brain were utterly destroyed' and 'the emotionless, inhuman quality of many lobotomised patients'. In general, therefore, he argues for an urgent application of much more systematic and forward-looking analyses of the social implications of advances in neuroscience than we have seen to date.

In particular, Blank shows much concern about the use of psychotropic drugs. As he expresses this worry:[15]

> Drugs are used to alter behavior because they are effective and convenient, not because of a compelling scientific consensus as to how they help patients.

So, in his view, drugs are sometimes used for non-medical reasons to achieve other policy objectives. He suggests that drugs are used as quick fixes for anxiety, depression and other problems, and illustrates his point by reference to the use of Ritalin for children and, more generally, to the protection of the individual rights of people who come under pressure to use drugs to make their behaviour more socially acceptable

So the point of interest here is that there is nothing new about advances in neuroscience raising difficult policy issues. Nor is there anything new in the lack of development of adequate policies to regulate the use of the new knowledge and novel technologies, despite much debate. As Blank wryly points out,[16] 'unlike brain policy, where there has been a dearth of action, there is a vast literature and considerable international scholarly activity in neuroethics'.

Whilst it is not the main focus of his book, Blank also, by drawing on the work of other scholars like Jonathan Moreno[17] who have investigated the military applications of modern neuroscience, applies his cautious approach to such applications in his seventh chapter, 'The Media, Commercial and Military Applications, and Public Policy'. For example, in a section on interrrogation he suggests that the recent public outrage over harsh treatment of suspected terrorists in the US 'has sparked interest in chemical approaches to interrogation'. He goes on to discuss the potential use of oxytocin in this context and then suggests that:[18]

> It is likely that there will be attempts to develop such substances, similar to the 1950s notion that LSD or other hallucinogens could be 'truth serums'.

Although, as he also points out, 'there is a consensus that physicians should not use drugs or other biological means' to take part in interrogations that are contrary to human rights and laws of war legal requirements.[19]

In his consideration of the use of neuropharmacology for national security Blank also looks at the use of, what he terms, calmatives and the use of chemical agents in the 2002 Moscow theatre siege. He notes the criticisms made of this use of fentanyl derivates by those concerned about the possible violation of the Chemical Weapons Convention but he suggests that this has not prevented continuing work on such less than lethal options and he references work on means of preventing respiratory depression when drugs such as fentanyl are used.

Blank is even more pessimistic about preventing the hostile misuse of biotechnology. Although he does not mention the Biological and Toxin Weapons Convention, he notes that:[20]

> Biotechnology's dual-use conundrum may hint at the difficulty of 'binning' advanced cognitive science research and development into offensive or defensive categories and may challenge traditional international security models.

So, with regard to the particular problem of developing means of controlling the applications of modern neuroscience for hostile purposes it would appear that Blank's general cautious optimism is transformed into a wary pessimism.

Yet, in his view, as modern neuroscience evolves and more applications of novel technologies appear, political debate is bound to increase and grow more intense[21] in regard to the issues he discusses in his chapter on the media and commercial applications and 'particularly with respect to military and national security applications'. The question here then is, what options may be available to strengthen the CBW non-proliferation regime against potentially dangerous developments in neuroscience?

Strengthening the regime?

Jonathan Moreno gave some consideration to the issues that are central here in his wide-ranging discussion of brain science and the military in the twenty-first century (see reference[17]). For example, he examined the implications of successful intranasal administration of oxytocin[22] and, in particular, he made an extensive review of debates in the United States this century in his chapter on non-lethal weapons.[23] I think it is evident from his account that the CWC and other international agreements, for example on human rights,[24] have had an impact in constraining proponents of incapacitating chemical and biological weapons in such debates.

Moreno, in fact, despite taking a realistic attitude to the possibility of misuse of the neurosciences, ends his book on a positive note:[25]

> We should be able to learn and apply the lessons of the new brain science for peaceful purposes. As the national security implications of neuroscience become more apparent, the pressing need to examine how our brains dispose us to peace as well as war should gain currency.

Indeed, he argues that the practical fields of conflict resolution and interventions in civil conflicts might benefit from our greater knowledge of the brain.

Moreno provided a positive preface to another recent study of the military implications of neuroscience,[26] in which he argued that the last ten years have seen a major increase in discussions of the ethical, legal and social implications of the neurosciences. There are certainly positive contributions in this study pointing to ways in which malign applications of the advances in neuroscience might be constrained, for example, Curtis Bell's account[27] of an attempt to set up a pledge for neuroscientists in a chapter titled, 'Why neuroscientists should take the pledge: A collective approach to the misuse of neuroscience'. Yet there is also plenty of cause for concern amongst the diverse eighteen chapters. For example, chapter 7 was contributed by Rachel Wierzman and the editor James Giordano[28] and is titled: '"NEURINT" and Neuroweapons: Neurotechnologies in National Intelligence and Defense'. This chapter states[29] that the major reports (discussed earlier here in Chapter 6) on the military applications of neuroscience were followed in the United States by 'a series of Strategic Multilayer Assessment (SMA) conferences' which 'considered the potential impact of neuroscientific understanding of aggression, decision-making and social behavior on policy and strategy pertaining to NSID [National Security, Intelligence and Defense] deterrence and influence campaigns'. So the issues, for example related to oxytocin, that were discussed at the end of Chapter 9 are still under review in the United States.

Wierzman and Giordano produced an updated and extended version of an earlier paper in their chapter of the study covering numerous issues, but the parts of most interest to us here are those dealing with neuroweapons in combat scenarios.[30] The headings of those parts of the chapter are shown in Table 12.1

In three detailed tables they set out the possibilities under each of the headings in Table 12.1 and, whilst the last heading leads into a discussion of the many drawbacks, it is clear that the authors' view is that the use of some of these weapons needs to be carefully considered. This is presumably in order to have an adequate defence if necessary, but no consideration is given in the chapter to either the CWC or the BTWC, nor does either convention have a mention in the book's index. Yet times are changing for life scientists, and scientists in associated areas of research, in the United States.

Table 12.1 Neuroweapons in combat scenarios

Neurotropic drugs
Neuromicrobiological agents
Neurotoxins
Practical considerations, limitations, and preparations

Source: Modified from Giordano, J. (2014) (Ed.) *Neurotechnology in National Security and Defense: Practical Considerations, Neuroethical Concerns.* Taylor and Francis Group, Boca Reton, CRC Press.

Legal developments in the United States

A Working Paper submitted by the United States for the July 2014 BTWC Meeting of Experts updated other States Parties on developments in the United States with regard to government oversight of life sciences dual-use research of concern. The paper recalled that the United States had issued a policy for the oversight of dual-use research of concern in 2012 that had placed requirements on federal departments and agencies (see Chapter 11) and announced that the government would shortly release a second policy that would expand DURC oversight to research institutions receiving US federal funding.

This policy, released on 24 September 2014, had numerous features of interest to anyone who supposed that modern societies would allow scientists to pursue any research they wished without oversight by government.[31] At first sight the policy, in requiring institutional oversight of dual-use research of concern, appears to cover only government departments and agencies, and institutions within and outside the United States that receive United States government funding. However, it also applies to US institutions that:[32]

> B ii. Conduct or sponsor research that involves one or more of the 15 agents or toxins listed in Section 6.2.1, *even if the research is not supported by USG [United States Government] funds*. [emphasis added]

Moreover, this section on the applicability of the policy ends by stating:

> Institutions that do not receive USG funds for life sciences research, but conduct life sciences research that has the potential to generate knowledge, information, products, or technologies that could be used in a manner that results in harm, are not subject to oversight as articulated in this Policy, *however, they are strongly encouraged to implement internal oversight procedures consistent with the culture of shared responsibility underpinning this Policy*. [emphasis added]

In short, all such institutions would be well advised to follow this policy.

The policy is also quite explicit in its requirement for education and training to be adequate for effective implementation of the policy. For example, the responsibilities of the principal investigators include:[33]

> E. Ensure that laboratory personnel (i.e., those under the supervision of laboratory leadership, including graduate students, postdoctoral fellows, research technicians, laboratory staff, and visiting scientists) conducting life sciences research with one or more of the agents listed in Section 6.2.1 of this Policy have received education and training on DURC.

And the institutional responsibilities include:[34]

> G. Provide education and training on DURC for individuals conducting life sciences research with one or more of the agents listed in Section 6.2.1 of this Policy, and maintain records of such education and training for the term of the research grant or contract plus three years after its completion.

Additionally, the responsibilities of the US Government,[35] set out in section 7.4, include to 'develop training tools and materials for use by the USG agencies and by institutions implementing this policy', and 'provide education and outreach to stakeholders about dual use policies and issues'.

Whilst these developments are further ahead than in other countries, there are certainly related developments taking place in other countries to consider what might best be done to ensure that the life sciences are protected from misuse. The seriousness and complexity of the problem became utterly clear on 17 October 2014 when the US Government halted further funding of gain-of-function research for a period in order that a deliberative review could be carried out. The announcement stated:[36]

> In the light of recent concerns regarding biosafety and biosecurity, effective immediately, the U.S. Government (USG) will pause new USG funding for gain-of-function research on influenza, MERS or SARS viruses, as defined

and:

> In parallel, we will encourage the currently-funded USG and non-USG-funded research community to join in adopting a voluntary pause on research that meets the stated definition.

During the one-year pause the NSABB would carry out the deliberative review and the National Research Council of the National Academies would convene a conference to review the NSABB draft recommendations. Then the NSABB would provide recommendations to the government.

International developments

Discussions of control measures for life sciences research are clearly also taking place in a number of other countries. These are likely to produce further national measures in some countries, but getting international agreement on the best way forward is likely to be a slower and more difficult task.[37]

In regard to the Chemical Weapons Convention, it is already clear that converting it from a primary focus on disarmament to a primary focus on non-proliferation – preventing the resurgence of chemical weapons around the world – is going to be a long-drawn-out process. For example, bringing facilities that produce by biosynthesis (OCPFs) fully under the verification system has been a clear *scientific* requirement for years, but it has not been possible to reach an agreement to do so. Yet perhaps the situation in regard to dealing with the potential loophole in Article II.9 (d) of the Convention may be becoming more tractable. Recently, a number of states have made it clear that they are not interested in incapacitating chemical agents for law enforcement and there are a number of different ways in which states can move towards an agreement on a restrictive interpretation of the meaning of this element of allowed peaceful purposes.[38]

Much will depend on the Eighth Review Conference in 2016 for the future of the Biological and Toxin Weapons Convention. Clearly, the new Intersessional Process agreed in 2011 has been far from a raging success in enabling joint actions to be agreed and implemented to strengthen the Convention.[39] As the Chair of the 2014 meetings noted in a letter to States Parties on 7 October 2014:[40]

> Our task now is to take this wealth of information and ideas [produced at the preceding Meeting of Experts] and consider how we might transform it into common understandings and effective action at the Meeting of States Parties.

He continued by noting that he had produced his own synthesis of what had been produced at the Meeting of Experts and suggested that further work should be undertaken before the Meeting of States Parties, in part to get greater clarity on 'where we might focus efforts on promoting effective action'.

The Chairman's synthesis paper included seven sections related to the SAI on advances in science and technology. They included, 'D. Voluntary codes of conduct and other measures to encourage responsible conduct' and 'E. Education and awareness-raising about risks and benefits of life sciences and biotechnology' but, most interestingly, under the final section, 'G. Any other science and technology developments relevant to the Convention', the paper pointedly noted that 'States Parties reiterated the value of continuing to consider, in future meetings, possible ways of establishing a more systematic and comprehensive means of review'. In short, we might say that they could go back and reconsider the excellent proposals made by several states at the Seventh Review Conference. If they cannot do so, continuing stagnation and potential disregard and disuse might be the fate of the Convention.

Conclusion

It seems reasonable to conclude that the jury is still out on Meselson's question of whether the biotechnology revolution will be applied in major ways to hostile purposes. It is not difficult to think of ways in which awful manipulation of the brain could result if humanity decides to go down that road.[41]

Yet travel down that road is not inevitable. We do not have to choose that route and scientists have a major role to play in protecting their work from such misuse. It bears repeating that scientists of considerable standing played significant roles in the decision of the United States to abandon its offensive biological weapons programme[42,43] and thus opened the door to negotiations in the BTWC (with its commitment to continue to work for negotiation of the CWC). Moreover, the BTWC negotiation was preceded by two major scientific reports,[44] one by the UN Secretary-General's Committee of Experts and another by a group of consultants to the World Health Organization.

Whilst it will not be sufficient, a necessary precondition for scientists to take on their increasing responsibilities as the revolution in the life sciences continues to gather pace in coming decades, is, as the UK Royal Society argued, for attention to be paid to ensuring that they have a better understanding of the security implications of their work. This point was again the first recommendation of what has been widely regarded as the most thorough recent analysis of the problem of dual use. Published by the German Ethics Council in late 2014 this recommendation stated that:[45]

> In view of the potential for misuse of dual use research in the life sciences, there is a need to increase the degree of awareness amongst the scientific community for these issues and to promote an underlying culture of responsibility.

So, a decade after the call for awareness and education of life scientists about biosecurity and the problem of dual use by the Fink Committee, they remain largely unachieved objectives.

The question that remains is, how long do we have to properly engage life scientists, like those studying the nervous system, to help work out how we best protect what they produce from large-scale hostile misuse? How long, in the present international situation, before we see the major use of chemical and biological weapons in the inevitable conflicts that will characterise the coming decades of the first half of the twenty-first century? What does seem certain in this centenary year of the start of the cataclysmic First World War is that there will be a prolonged period of instability[46] and that, so far, we have not been overly successful in dealing with these conflicts.[47]

Reflecting on the numerous books that have appeared recently, which try to explain how the First World War came about, Lawrence Freedman, Professor of War Studies at Kings College, London, concluded[48] that there were no sure lessons, but decision makers always have choices and that they should make their choices with the best possible information and scepticism about military plans. Obviously, had they better understood the nature of the warfare that would be possible with the weapons available the decisions to go to war would have been taken with much more caution in 1914. Thus today neuroscientists with a clear grasp of biosecurity and the problem of dual-use, and the professional organisations to which they belong, surely have many roles to play in helping to prevent the proliferation and potential use of novel chemical and biological weapons. These include not only being careful about the research they do and what they publish, but also following the efforts to strengthen the national and international policies and regimes designed to prevent the misuse of their work, helping inform the public and policy makers of both the dangers and the potential benefits of their work, and ensuring that their students are well-informed and engaged in this effort during their working lives because the problem of biosecurity and dual use will not be resolved for decades to come. Above all, it needs to be understood that it will be too late to act if the use of novel neuroweapons becomes widespread and commonplace as a method of warfare and terrorism. The effective long-term governance of neuroscience will depend in good part on the continued effective engagement of well-educated scientists with the public, media, military and politicians well into the middle years of this century.

References

1. Meselson, M. (2000) Averting the hostile exploitation of biotechnology. *The Chemical and Biological Conventions Bulletin*, **48**, 16–19.
2. Evans, R. J. (1997) *In Defence of History*. London: Granta Publications. pp. 60–61.
3. MacKenzie, D. (2014) Imagine there's no countries. *New Scientist*, 6 September, 30–37.
4. ibid, p.36.
5. Kissinger, H. (2014) The World in Flames. *The Sunday Times*, 31 August, News Review pp.1–3.
6. Development, Concepts and Doctrine Centre (2010) *Global Strategic Trends – Out to 2040*. Ministry of Defence, London.
7. ibid, p.10.
8. ibid.
9. ibid, p.14.
10. ibid, p.15.
11. See, for example, Bardin, J. (2012) From Bench to Bunker: How a 1960s discovery in neuroscience spawned a military project. *The Chronicle of Higher Education*, 9 July. Available at <http://chronicle.com/article/From-Bench-to-Bunker/132743>. 19 September 2012.
12. Blank, R. H. (1999) *Brain Policy: How the New Neuroscience Will Change Our Lives and Our Politics*. Washington, D.C: Georgetown University Press.
13. Blank, R. H. (2013) *Intervention in the Brain: Politics, Policy, and Ethics*. Cambridge, MA: MIT Press.
14. ibid, p.36.
15. ibid, p.46.
16. ibid, p.65.
17. Moreno, J. D. (2006) *Mind Wars: Brain Research and National Defense*. New York: Dana Press.
18. Reference 13, p. 224.
19. ibid, p. 226.
20. ibid, p. 223.
21. ibid, p. 227.
22. Reference 17, pp. 90–91.
23. ibid, pp. 163–184.
24. Crowley, M. J. A. and Dando, M. R. (2015) *The Use of Incapacitating Chemical Agent Weapons in Law Enforcement*, in press.
25. Reference 17, p. 204.

26. Giordano, J. (2014) (Ed.) *Neurotechnology in National Security and Defense: Practical Considerations, Neuroethical Concerns*. Taylor and Francis Group, Boca Reton: CRC Press.
27. ibid, pp. 227–238.
28. ibid, pp. 79–114.
29. ibid, p. 80.
30. ibid, pp. 96–109.
31. United States (2014) *The United States of America Government Policy for Institutional Oversight of Life Sciences Dual Use Research of Concern*. Available at <http://www.phe.gov/s3/dualuse>. 20 October 2014.
32. ibid, p. 8.
33. ibid, p. 12.
34. ibid, p. 16.
35. ibid, p. 18.
36. United States (2014) *U.S. Government Gain-of-Function Deliberative Process and Research Funding Pause in Selected Gain-of-Function Research Involving Influenza, MERS, and SARS Viruses*. Available at <http://www.whitehouse.gov/blog/2014/10/17/doing-diligence-assess-risks-and-benefits-life-sciences-gain-function-research>. 20 October 2014.
37. Netherlands National Academy (2014) *Report of a Debate on Gain-of-Function Research between Professor Giorgio Palu and Professor Simon Wain-Hobson*, Amsterdam, 25 June.
38. Crowley, M. J. A. and Dando, M. R. (2014) *Down the Slippery Slope? A Study of Contemporary Dual-use Chemical and Life Science Research Potentially Applicable to the Development of Incapacitating Chemical Agent Weapons*. Policy Paper 8, Biochemical Security 2030 Project, University of Bath, November.
39. Dando, M. R. (2014) *To What Extent Was the Review of Science and Technology Made More Effective and Efficient at the 2013 Meeting of BTWC States?* Policy Paper 5, Biochemical Security 2030 Project, University of Bath, May.
40. Ambassador Urs Schmid (2014) *Biological Weapons Convention Meeting of States Parties*. BWC Implementation Support Unit, United Nations, Geneva, 7 October.
41. Serronia, M. I. J. (2007) Awakenings (1990): The epidemic of children who fell asleep. *Journal of Medicine and Movies*, **3**, 102–112.
42. Tucker, J. B. and Mahan, E. R. (2009) *President Nixon's Decision to Renounce the U.S. Offensive Biological Weapons Program*. Center for the Study of Weapons of Mass Destruction, Case Study 1, National Defense University, Washington, DC, October.
43. Reference 42, p. 3. Footnote 9 lists some of the scientists involved such as 'Harvard molecular biologist Matthew Meselson'.
44. Reference 42, p. 7.
45. Deutscher Ethikrat (2014) *Opinion: Biosecurity – Freedom and Responsibility of Research*. German Ethics Council, Berlin. (p. 179).
46. Hass, R. N. (2014) The unravelling: how to respond to a disordered world. *Foreign Affairs*, November/ December, 70–79.
47. Boot, M. (2014) More small wars: counterinsurgency is here to stay. *Foreign Affairs*, November/December, 5–14.
48. Freedman, L. D. (2014) The war that didn't end all wars: what started in 1914 – and why it lasted so long. *Foreign Affairs*, November/December, 148–153.

Index